Emotional Selection

How Your DREAMS Evolve Your MIND

RICHARD COUTTS

WONDERS
PUBLISHING

www.wonderspublishing.com

This book contains advice and information relating to mental health care. It should not be used to replace the advice of trained, mental health care professionals. If you suspect you have a mental health problem, it is recommended that you seek the help of mental health care professionals. The publisher and author disclaim liability for outcomes that may occur as a result of applying the methods suggested in this book.

Ensuring the accuracy of the information in this book is a top priority for the author. To report errors or to provide suggestions, please visit www.emotionalselection.com.

Text set in Caslon using LaTeX and the memoir class. E-book versions were generated by tex2ebook and edited with Calibre. Cover design by Richard Coutts.

1 1 0 K S J 0 0 0 0 0

Coutts, Richard. *Emotional Selection: How Your Dreams Evolve Your Mind*
Includes bibliographical references and index
1. Dreaming. 2. Psychology. 3. Systems Engineering.

ISBN 978-0-578-45447-4

for Emma and Melanie

Table of Contents

Part Two: Your Dreams

Appendices

Dream Theme Sections

Emotional Selection

Introduction

> *Why does the eye see a thing more clearly in dreams than the imagination when awake?*
>
> — Leonardo Da Vinci

To explore the impact of dreaming on our daily lives, I designed a simple study to test whether people who frequently dream about romantic or sexual attraction are more likely to be in a relationship in the future, compared to those more apt to dream about problems at work, flying, falling, being attacked, or other dream themes.

For my study population, I needed participants who were seeking a relationship. Where does one go to find people searching for love? To an online dating website, of course. Plenty of Fish generously sponsored my study by hosting an invitation to my survey. Over fifteen thousand visitors clicked on the invitation, entered a recent dream, and answered questions about their dating activities.

Many participants had emotional dreams about ex-partners:

> *I was chatting and getting on with my ex … I felt totally healed from all the heartbreak. There was nothing but complete serenity, calm, and comfort.*
>
> *I dreamed that my ex-boyfriend had sex with one of my best girlfriends. He did this out of spite, to the extent that he sent me pictures of the act. The images were as clear as day. I woke up in the sweats with tears rolling down my face.*

I dreamed that my ex-girlfriend was in bed with me, and I woke up talking to her when she wasn't there.

Some had pleasant dreams about future partners, real and fictional:

All I remember [was] it was about me and a guy I have never met, but I felt so comfortable around him and a lot of love toward him.

My friend, (a girl I will probably be dating in the near future) and I were at the airport getting ready to take a trip somewhere. We were sitting at the airport, just joking around having a good time.

I met a girl I had never met, who knew me already. It was pleasant and friendly—nothing uncomfortable at all. She came to me and held my hands.

Some had sexual dreams:

I was with my ex-girlfriend, making love to her on the beach.

I was in a motel and a friend came to see me and brought over some more friends. I started liking this one man and, from what I can remember, he and I ended up making love, and what he was doing to me I haven't had in years.

I met a girl at a lake for a picnic. We never met before. It was our first meeting. We got along great. We had wine and food and, as the night went on, we made love under the stars.

I was with an old friend (male) whom I haven't seen in 25 years, and we were finally having the reunion we'd both hoped for. Lots of smiling, laughing. We were eating for a while, then I saw us back at my hotel room and we were for the first time making love. It felt so very real.

Many participants dreamed of themes unrelated to attraction. Work, missing travel connections, meeting famous people, and being chased were common dream themes:

Zombies [were] taking over the world. [A stranger] and myself [were] driving cross country, picking up survivors. [We] saved a baby and focused most of our attention on keeping her alive.

I was at work and everybody was acting so differently toward the end of the dream; people were not being nice to me and I was incredibly confused.

I was talking to President Lincoln, and he was asking me why it had taken me so long to get back to him. The conversation turned to my divorce and the troubles of a broken marriage. He gave me some advice, then we proceeded to talk about social issues.

Were the participants who dreamed about attraction more likely to be in a future relationship? Yes. Three and twelve months later, over one thousand participants completed follow-up surveys that updated their relationship status. Singles who dreamed of attraction toward an ex-partner were a whopping 83 percent more likely to be in a future relationship.[1] Meanwhile, those who did not dream of attraction, or who dreamed of attraction to someone they did not know intimately, such as a stranger, acquaintance, or celebrity, were more likely to remain single.

Why were dreams about an ex-partner better predictors of being in a future relationship than dreams about an attractive non-partner?

Because dreams about an ex were more likely to include an emotional theme, such as love or heartbreak, while dreams about an attractive non-partner were more likely to include a sexual theme. So, while a dream of hooking up with a sexy celebrity may be enjoyable, an emotional dream of navigating a relationship is a better predictor of being in a future relationship.

The benefits of dreaming about relationships also applied to participants already in a relationship. Over two hundred of the survey respondents reported as being in a relationship at the start of the study. These partnered participants were more likely to stay in their relationship if they dreamed of attraction to a current or former partner. Meanwhile, those who did not dream of attraction, or who dreamed of attraction to a person who was not their current or former partner, were more likely to break up. In other words, for many of my study participants, their dreams of attraction literally came true.

There are always exceptions in studies like this. For instance, several singles who did not dream about a relationship still entered a relationship in the future, and many initially partnered participants remained in their relationship despite dreaming about work or another non-relationship theme. However, the data suggest that our dreams profoundly contribute to our ability to subsequently meet our needs. Fortunately, for readers whose dreams do not align with their imagined futures, we will discuss proactive steps you can take while awake to guide the content of your dreams, thereby influencing your future.

To be clear, dreams improve more than just your personal relationships. As we will discuss, dreams improve your ability to meet all your human needs, from your most primitive needs for food and shelter to your higher, cognitive needs for aesthetics and self-fulfillment.

My findings, as well as those of other researchers that we will discuss, suggest that dreams serve a function, but what is that exact function?

A Theory

According to my theory, while we sleep, our mind uses a two-phased process to evolve our understanding of ourselves and our worlds. During the first phase, which I call the accommodation phase (for reasons I will explain in Chapter One), our mind incorporates dreams to improve our ability to navigate waking life. For instance, we may dream of defending ourselves and others, being loved and loving, working, being creative, surrounding ourselves with beauty, or reaching our potential. Regardless of our dream, our mind tentatively adopts its theme to create an altered perspective of our world and our place in it.[2]

Because the mind is an extraordinarily complex system, alterations can introduce accidental conflicts. Therefore, our mind enters a second phase in which dreams serve as test scenarios to evaluate prior changes. During these latter dream tests, changes that ease anxiety or sadness, or otherwise appear emotionally adaptive, are adopted. Conversely, changes that trigger anxiety or are inferior in other ways are discarded. Because emotions play a critical role in selecting adaptive changes, I call this second phase (and my theory in general) emotional selection.

You may notice similarities between emotional selection and another theory. Charles Darwin also used two phases to describe an evolutionary process. His first phase, which he called *variation*, introduces changes to the characteristics of offspring. The beneficial changes help offspring survive, while the detrimental changes make offspring less competitive. Darwin's second phase, *natural selection*, selects the offspring with beneficial characteristics. After the natural selection phase, the process returns to the variation phase and begins anew. With each generation, the two phases of variation followed by natural selection evolve a species.

Natural selection and emotional selection both employ a two-phase process to evolve complex systems, but their similarities end there. Natural selection evolves the genome. Emotional selection evolves your mind through dreams.

Can emotional selection be used to interpret dreams? Yes, though I prefer the term "deconstruction" over "interpretation." Interpretation requires translation, but understanding dreams requires no such step. The purpose of dreams is simple: they are constructs that modify and test your mind. To understand the role of a particular dream, you deconstruct it. For instance, the participants of my study deconstructed their dreams by answering a few simple questions, including whether they were attracted to anyone in their dream.

We have many needs in our lives, so we have many types of dream themes to help meet those needs. By reconsidering our dreams, we can determine which needs our mind is attempting to address during sleep. This approach to dream deconstruction can be applied to virtually any dream, including universal themes such as missing travel connections, meeting celebrities, being lost, facing natural disasters, being unable to find a toilet, enduring malfunctions, being unprepared for a test, teeth falling out, hair falling out, visiting with past loved ones, falling, flying, and more. As we will discuss, deconstructing dreams can provide insights into the areas of our lives that our minds work to improve.

Describing dreams as mental tests within the framework of emotional selection also explains why dreams appear so bizarre. Imagine your neighbors intentionally crashing their automobiles into walls—you would describe their behavior as bizarre. However, when engineers perform the same act during crashworthiness tests, their acts are rational—because they are tests. Similarly, dreams appear bizarre until you consider them as tests. Take the common dream of being accidentally naked in public. This dream seems bizarre until you consider it to be a test of your ability to cope with humiliation. Likewise, we can deconstruct virtually any bizarre dream to identify the rational test performed on your mind.

Emotional selection improves our ability to meet our needs. However, while awake, we must also cope with instances when our needs are unfulfilled. Emotional selection aids with this, too. Our dreams often include themes of unmet safety needs as we succumb to an attacker, unmet love needs as an ex rejects us, or unmet esteem needs as we struggle at work. Such dreams of unmet needs modify and test our ability to pick ourselves up and try again.

From childhood through adulthood, our mind continuously evolves to meet the ever-shifting demands of daily life. As we will discuss throughout this book, dreaming is an essential part of this mental evolution.

The Parts

This book is divided into two parts. *Part One: The Theory* delves into how emotional selection relates to sleep, the mind, and your waking life. Here, we will discuss several characteristics of dreaming, such as why dreams are difficult to remember and why dreams often recur. Part One concludes with a comparison of emotional selection to other contemporary and historic dream theories.

In *Part Two: Your Dreams*, we will explore how to deconstruct your personal dreams. Using these insights, we will investigate how to influence your dreams to improve your mind's ability to meet your needs in the future.

Emotional Selection: How Your Dreams Evolve Your Mind elucidates the role dreams played in shaping the person you are today, and provides tools you can use to shape the person you will be tomorrow.

Part One

The Theory

ONE

Bizarre

A White Rabbit with pink eyes ran close by her. There was nothing so very remarkable in that; nor did Alice think it so very much out of the way to hear the Rabbit say to itself, "Oh dear! Oh dear! I shall be late!" (When she thought it over afterward, it occurred to her that she ought to have wondered at this, but at the time it all seemed quite natural.)

— Lewis Carroll,
Alice's Adventures in Wonderland

DREAMS are famously bizarre. Imagine yourself in the following situation:

You are on an airliner in flight. You don't know how you got there, but somehow that is immaterial.

"Where are we going?" you ask, turning to the person seated next to you. No response. You notice his eyes are closed, so you assume he is napping, but upon closer inspection, you see he is made of plastic, like a manikin or life-sized doll.

With no flight attendant in sight, you unbuckle your seatbelt and head for the cockpit. "Excuse me, Captain," you say to the back of the pilot's head, "where are we ..." The view through the cockpit window arrests you. The plane is on final approach and about to touch down, but at no airport like you have ever seen before. It is in the desert, with no city in sight. There is a runway, but no planes or control tower—only vehicles and equipment scattered about.

With no time to return to your seat, you brace for landing, but the moment passes. No touchdown, no sound, nothing. The plane simply keeps descending. Suddenly, the left wing hits the ground. You reach for a grab bar, but come up short as the fuselage slams to the runway, sending you to the floor and the plane into an uncontrollable skid.

The roar of the aluminum belly grinding against the runway announces that the plane's landing gear was undeployed. As you scramble to your feet, your attention once again goes out the cockpit window to what appears to be towering steel pylons strewn across the airplane's path. "Who put those THINGS in the middle of the runway?" is the one thought you muster, as though everything up to this moment has been perfectly rational. You brace for another impact.

The second impact is worse than the first, shearing off the wings while hurling you into the instrument panel. You regain consciousness as a massive fireball enters the cockpit.

Apologies. If you thought the above dream-like scenario was a dream, then I misled you. It was an actual event, although fortunately neither you nor anyone else was injured. It was a test in which NASA remotely piloted a jet airliner and intentionally crashed it in the desert. The only "people" onboard were crash test dummies.

Why did I insert you into a crash test conducted by NASA? To demonstrate the outward bizarreness of tests. During tests, engineers crash, shake, trounce, and otherwise subject designs to extremes. While taxing, such rigors ensure designs function as intended. Testing may appear surreal to the uninformed onlooker, but it is essential for evaluating design modifications.

As we will discuss throughout this book, many of your dreams are tests, which is why they often seem bizarre. Similar to how engineers employ tests to evaluate changes to their designs, the mind employs dreams to evaluate its changes. More on this later. Let us return to our airline crash test to gain insight into testing and dreaming (and learn why NASA decided their experiment was worth destroying an expensive airliner).

In the early 1980s, NASA attempted to improve the survivability of airline crashes. After extensive research, they developed FM-9—a fuel additive that reduced the tendency of disturbed fuel to form fireballs. Preliminary tests showed it had potential, but there was only one way to be certain. On December 1, 1984, NASA conducted a "controlled impact demonstration." With its fuel tanks filled with jet fuel and FM-9, a Boeing 720, equipped with cameras, sensors, and test equipment, was intentionally crashed on a makeshift runway in the desert. To ensure substantial damage, NASA kept its landing gear up and installed steel "wing cutters" on the runway.

Unfortunately, the test results were not good. An immense fireball engulfed the plane that took firefighters an hour to extinguish. Although some researchers argued that FM-9 reduced the size of the fireball, NASA abandoned the project due to the disappointing test results.

I placed you within the controlled impact demonstration to show how strange it would feel to unknowingly experience NASA's test. Intentionally undeployed landing gear, crash test dummies for passengers, and steel wing cutters would certainly be bizarre if used during a commercial flight, but they created the necessary conditions for a successful crashworthiness test.

According to emotional selection, the mind faces challenges similar to those of engineers: ideally, changes to the mind work as intended, but often they fail. So, your mind tests its changes with dreams. Similar to how the NASA engineers used steel pylons, your mind often conjures unusual objects during dreams to test itself, which is why you may find yourself with superpowers, back in high school, or face-to-face with a lion in your dreams. Such oddities facilitate your dream tests.

A healthy skeptic may question whether tests performed by engineers equate to those performed by the mind during sleep. After all, what properties does an airplane share with the mind? On the surface, there are very few. However, as we will discuss, airplanes, computers, automobiles, nature, the universe, and the mind share one thing in common: they are all systems. And, like most systems, they evolve. Airplanes and automobiles evolved from contraptions that one or two people could build in a garage to the complex, computer-controlled systems we know today; computers evolved from rudimentary programmable calculators, composed of tubes and wires, to artificially intelligent supercomputers; the universe evolved from the big bang to its many systems, including galaxies and our solar system; and the capabilities of our minds evolved from the rudimentary understanding we held as infants to a system that comprehends the nuances of complex emotional and intellectual situations. Systems surround us and comprise us, and these systems continuously evolve.

Mental Structures

For a deeper understanding of dreaming, we need a description of some of the workings of the mind. Fortunately for our discussion, the philosopher Immanuel Kant provided insights into how the mind organizes the understanding that our dreams modify.

In the late 18th century, Kant reflected on how the mind assigns meaning to our surroundings. He understood that vision converts light into images and hearing converts vibrations into sounds, but how does the mind interpret the images to mean "friend" or "book," or sounds to mean "barking dog" or "thunder"? His musings led him to a simple idea—the mind maintains mental structures that hold concepts.

Kant named his structure "schema" (from the Greek *skhēma* meaning "shape") and described how schemas interconnect to form a network that contains our understanding. So, the schema "sphere" connects to "ball," which connects with the schemas "roll" and "bounce." Such simple schemas connect to more abstract schemas, such as "play" and "sport." While individual schemas represent simple concepts, when networked together, they form complex understandings.[3]

Schemas also represent intangible concepts, including the motives and intentions of people in our lives. As we interact with strangers, friends, and loved ones, our schemas continuously assign meaning to their behaviors. Facial expressions, such as winks, smiles, sideways glances, and turned-up noses, are all characterized by our schemas.

In addition to describing our surroundings and the people in our lives, our schemas also represent ourselves, including our self-perceptions of likeability, competence, shyness, attractiveness, and intelligence. The network of mental schemas in our mind describes everything we hold to be true.

Schemas profoundly impact our behavior. For instance, those that support competence can eliminate self-doubt, allowing us to think with clarity and solve problems under pressure; schemas that bolster a self-image of athleticism help us maintain focus during competition; those that reinforce our self-perception of talent can calm our nerves during performances; and schemas that promote a self-image of likability enhance our confidence, aiding us in making new friends and finding a life partner. In short, schemas that represent a healthy self-image make us feel successful, worthy, and lovable.

Our schemas continually resolve conflicts during our daily lives. We strive for independence yet find solace in the company of others. We are sexually attracted to many people, yet seek loving, monogamous relationships. We prioritize our personal safety yet quickly jeopardize it to help those in need. These are a few of the countless drives that directly conflict. Schemas produce the perceptions that help us navigate the complex, ambiguous, and contradictory landscape of life.

As we mature during our lifetimes, our schemas evolve. As children, our biggest concerns include play and safety. As we mature into adulthood, our priorities shift to jobs, family, and more. As we assume responsibilities, our complex network of schemas evolves to meet life's shifting demands. Despite the universality of schemas, no two schema networks are identical. Your unique schemas differ from those of your parents, friends, and acquaintances.

Kant's ideas proved so appealing that today they help unify the brain sciences. The fields of psychology, psychiatry, and neurology all emphasize the importance of schemas.

Psychologist Jean Piaget expanded on Kant's elegant model in the early 20th century by describing how schemas continually evaluate and adapt to the information they represent. When existing schemas interpret an event, Piaget referred to this process as *assimilation.* For instance, a child who encounters a brown dog in the park assimilates the animal using the schemas "brown" and "dog." After seeing a horse for the first time, the child may declare, "Big dog!" After the parents inform the child that the "big dog" is actually a different kind of animal, the child's mind forms new schemas to support the concept of "horse," a process Piaget called *accommodation.*

As children mature into adulthood, their schema networks grow in size and intricacy as they accommodate new information. As adults assume new roles and experience new environments, their schemas also adapt through accommodation. This daily evolution of schemas explains our expanding understanding of our surroundings and ourselves.

According to emotional selection theory, while you sleep, your mind processes dreams that schemas accommodate (which is why I call this phase the "accommodation" phase). As an example, schemas may elevate a person's perceived independence during this phase by accommodating dream scenarios in which the dreamer successfully acts independently. These dreams may include themes of performing a task admirably before a crowd or executing a mundane daily chore in solitude. Regardless of the theme, at the close of the accommodation phase, schemas hold a slightly adjusted understanding.

Because schemas form a complex network, even seemingly benign changes can be maladaptive. For instance, improving a person's perceived independence may appear universally beneficial, but schema networks also support self-perceptions that conflict. Toddlers with overdeveloped perceptions of independence may wander into dangerous situations, which is why they benefit from perceived dependence rather than independence. So, schemas balance perceptions of independence with dependence.

After the accommodation phase, the emotional selection process enters a second phase in which dreams serve as test scenarios for evaluating tentative prior accommodations. During these dream tests, new accommodations that relieve anxiety, sadness, or otherwise appear beneficial are selected, while accommodations that exacerbate emotional responses or otherwise compare negatively to existing schemas are culled.

For a toddler who dreams of an elevated sense of independence, his or her mind may test these changes with a dream in which the child navigates a gathering of people without a caregiver. During the dream tests, the child's mind monitors the emotions triggered by the dream themes. These emotions can be negative, such as stress or sadness, or positive, such as excitement or pride. After the dream tests, emotional selection retains the new accommodations if the dream tests suggest they are emotionally adaptive and will help the child thrive. Otherwise, the accommodations are culled.

Which schemas do dreams modify and test? Not those that describe unambiguous concepts such as "book," "table," and "automobile." These schemas are relatively straightforward, so they do not require testing with dreams. Schemas that characterize ourselves and others, on the other hand, are riddled with ambiguities. Is a passing comment by an acquaintance intended as playful teasing, or was it meant as biting sarcasm? Is a workplace mistake comically embarrassing or utterly humiliating? Our dreams target complex schemas for modification and testing because complex schemas are difficult to adapt.

As mentioned, both emotional selection and natural selection employ two phases to evolve a complex system. However, they work on very different time scales. The natural selection cycle repeats once every generation—you inherited genes from your parents, who likewise inherited genes from their parents, and so forth. The natural selection process is slow, having evolved our brain with each passing generation over millions of years. Conversely, the emotional selection cycle repeats several times during a typical sleep. In other words, natural selection contributed to the schemas with which you were born, while emotional selection evolves your schemas during your lifetime.

The costs of emotional selection are far less than those of natural selection. According to natural selection, offspring unfit for their environment are more likely to die to make room for those better adapted. In contrast to the life-and-death struggles of natural selection, emotional selection employs less perilous tests: dreams.

Discussion

This chapter outlined emotional selection's description of dreams as modifiers and tests of mental schemas. The balance of this book explores the evidence supporting the theory of emotional selection, how it provides insight into the workings of the mind, and how these insights help you understand your dreams and yourself. We continue our exploration in the next chapter with a discussion of the surprising relationship between the sleep cycle and the cycle of accommodation and emotional selection.

Shall we continue assimilating and accommodating?

TWO

Offline

It is a common experience that a problem difficult at night is resolved in the morning after the committee of sleep has worked on it.

—John Steinbeck

IN 1951, Eugene Aserinsky, a doctoral candidate, was working on his dissertation in sleep research. Lacking funding for his experiments, he used an old electroencephalogram (EEG) and conscripted his eight-year-old son, Armond. After attaching electrodes to his child's scalp, Aserinsky instructed the boy to look left, right, up, and down as the EEG needles inked the brain activity of the child's responses onto scrolling paper. The 29-year-old Aserinsky then put his son to bed, as he had done countless times before, but this time on a cot in his research laboratory. As slumber overtook the boy, the needles of the EEG likewise dozed, reaching a low-frequency rhythm that recorded his son's oncoming sleep. With little to do but nibble on pretzels and ply himself with coffee to stay awake, Aserinsky hunkered down for an expected long night of watching and waiting.

Soon afterward, something seemingly went wrong. The needles sped up and danced again, scratching back and forth as they had before the child fell asleep. Disappointed that his son apparently awoke, the researcher approached his child. To his surprise, the boy was fast asleep. Aserinsky glanced back and forth between his instruments and his sleeping son, presumably bewildered and excited. He was the first person to witness high-brain activity during sleep.

At that time, Aserinsky's mentor, Nathaniel Kleitman, was the world's only full-time sleep researcher. Sharing his student's enthusiasm for the discovery, Kleitman organized a full-fledged study. Kleitman, a full professor with the associated resources, did not need to rely on family members for experiments.

After recruiting twenty adult participants and monitoring their brain activity during sleep, the pair discovered that the brain cycled in and out of high brain activity several times during a typical night's sleep. The researchers named this high brain activity rapid eye movement (REM) sleep and found that people woken during REM sleep were more likely to report a dream than those woken during non-REM sleep. After two years of research, Kleitman and Aserinsky published their results in a modest two-page paper that became the cornerstone of modern dream research.

Sleep Stages

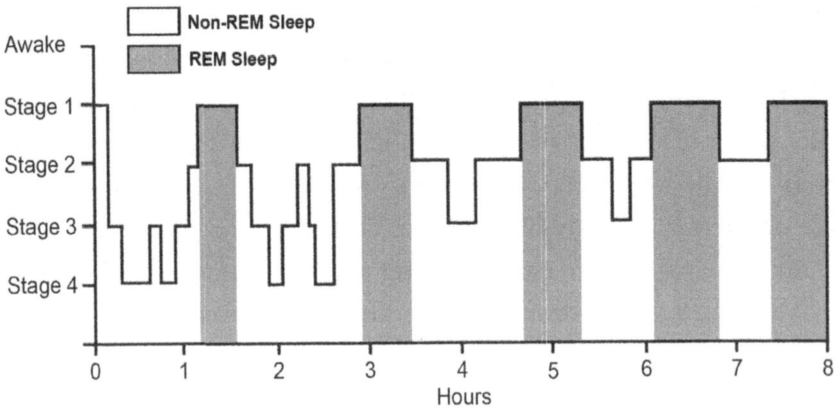

The above plot depicts brain activity during a typical eight-hour sleep. As shown, researchers divide this brain activity into four stages: Stages 3 and 4 are the deepest and collectively called *slow-wave sleep*

(because of their low-frequency brain waves). As brain activity increases, sleep passes through Stages 2 and 1 and enters its highest-frequency stage, REM sleep. When REM sleep finishes, brain activity slows, and the cycle repeats.

While dreams are often associated with REM sleep, people also dream during non-REM sleep. However, REM dreams and non-REM dreams typically differ. REM dreams tend to be more extreme and bizarre than non-REM dreams, having themes such as flying or teeth falling out. Conversely, people woken during non-REM sleep typically report either no dream or a mundane dream, such as reading or doing housework.

As shown in the previous plot, non-REM sleep is longest and deepest during the first cycle and becomes shorter and shallower with later cycles. Meanwhile, REM sleep is shortest during the first cycle and lengthens throughout the night. Also, non-REM sleep is typically the first cycle, and REM sleep is typically the last cycle when you wake naturally, as opposed to being roused.[4] Those occasions when you wake from deep non-REM sleep are those groggy "Where am I?" awakenings.

The plot of the sleep cycle resembles a theoretical plot of the modification and testing phases used to evolve systems:

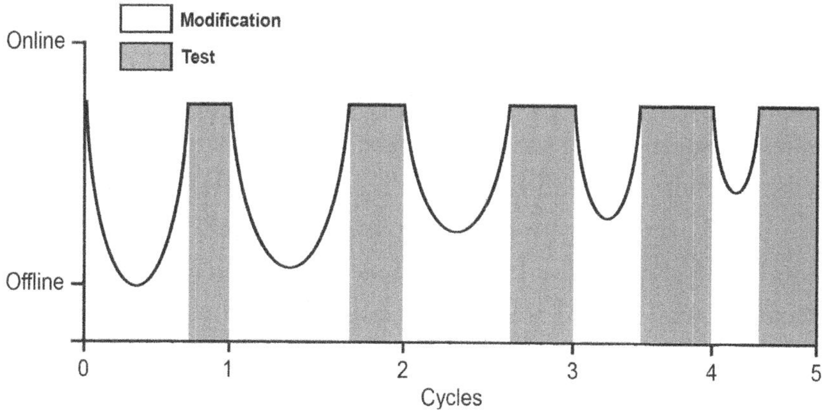

As shown in the above plot, the deepest modifications to systems typically occur during the early stages, in which systems are "gutted" or engineers "go back to the drawing board." These early modification phases typically introduce many defects. For instance, if the system is a software program, it may be buggy. If it is an engine, it may run roughly. Regardless of the system, finding defects during these early phases is typically easy. So, early testing phases tend to be short.

Once defects are identified and isolated, the process circles back to the modification phase. With the completion of each cycle, the robustness of the system improves, resulting in progressively shallower and shorter modification phases. Conversely, the testing phases become longer as emphasis shifts from modification to testing to ensure the system works as intended. Because emotional selection also comprises a modification phase (accommodation) and a testing phase (emotional selection), a plot of the emotional selection cycle shares the same characteristics as a plot of the cycle of modification and testing.

According to emotional selection theory, the similarities between the shape of the sleep cycles and the shape of the cycles used to evolve complex systems are not coincidental—non-REM sleep is when the accommodation phase occurs, and REM sleep is when the emotional selection phase occurs. So, after falling asleep, emotional selection sets to work improving your schemas during non-REM sleep by accommodating the themes of dreams. These early schema modifications are the deepest, and therefore, the most likely to be conflict-ridden. Sleep then transitions to REM sleep, in which dreams test the newly minted accommodations. As sleep progresses, non-REM sleep cycles become shallower as schema modifications decrease, while REM sleep cycles lengthen as dream tests increase. So, as shown in the figure below, we can update the plot of the sleep cycles with labels that show the relationship between the cycles of sleep and the cycles of emotional selection:

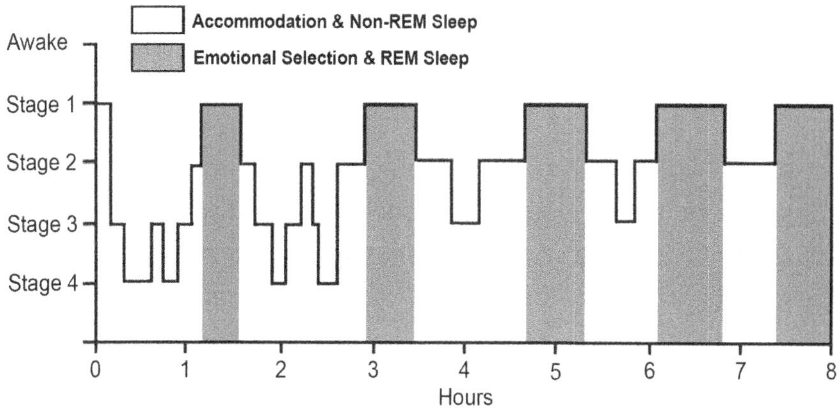

Researchers have observed that REM dream themes grow increasingly surreal throughout the night, with REM dreams just prior to waking being the most extreme and bizarre of all.[5] This sequence is ideal for testing schemas. Early, non-bizarre dreams gently test initial accommodations, which are often riddled with conflict. As sleep progresses and morning nears, emotional selection changes emphasis from creating accommodations to testing them to ensure they work as intended. So, we experience a blizzard of bizarre dreams before waking as our mind taxes its new accommodations to confirm they are ready to begin a new day.

Modifying and testing the same schemas in a single sleep requires dreams with similar themes across sleep cycles. Researcher Rosalind Cartwright often observed this pattern:

> Any group of dreams collected on one laboratory night from the same person makes such a coherent statement that it is hard to entertain the idea that these are ran-

> dom events. The dreams of a night constitute a body of an interrelated whole. As a group, they reflect the dreamer's personal response to waking life and relate these responses to the remembered past and to the hopes and fears of the future. What is more, they do this regularly each night throughout [their lives].[6]

In addition to dreams maintaining a common theme throughout the night, those that occur early in sleep often reflect recent events, while later dreams often tap older memories.[7] Cartwright witnessed this pattern in the dreams of Jerry. After the electrodes of an EEG were attached to his head, Jerry fell asleep and dreamed:

> *I was playing with the leads from these electrodes in my mind, and trying to figure out where they went. I was pulling out plugs and putting others back in there. It didn't seem to be any real order to it. I can't really remember a whole lot. I know I played around for a while. I stuck the leads in once and they didn't seem to be right, so I put them in a different way. I do remember seeing a picture and I had them all stuck in a little box down at this end.* [You saw a picture?] *Well, in a mental image, what the terminal box looked like that I stuck the leads into.*[8]

Cartwright woke Jerry five times during the night to record his dreams. In his next dream, Jerry drove to the clinic and encountered a confusing intersection where the stop sign was turned around. After negotiating the intersection, he found himself lost, unable to locate the clinic. His third dream was similar, but more abstract. In it, he drove to the library to return a book. After arriving, he entered the library but did not know where to return the book. His fourth, pleasant dream took him back to his childhood, in which he made mud pies with a little girl. His fifth and last dream regressed further, in which he was four years old and sat in a highchair with his mother washing his face.

Jerry was a first-year medical student who Cartwright described as a bright, "tall, rangy, farm boy from downstate Illinois [who was] rather shy and unsophisticated." Some weeks later, Jerry reported to Cartwright his personal interpretation of the dreams, which Cartwright paraphrased:

> They all followed from that first one about trying to figure things out for myself and not being sure I can do it right. This place (medical school) really has me working to keep up. I'm just a country boy, and I've never lived in a big city before. Guess I've always been like that, though. My father always gave me impossible jobs to do, like keeping a second-hand car running for him. If I couldn't fix something, boy, he really blew up. He would never buy new parts but always had three or four old junkers around, and if I needed something, I had to rob the others to keep one going. It was impossible. I had to keep trying, but I could never win. I thought I left that frustration all behind me with the hicks and the turkey farmers, but I guess I haven't because I still get mad talking about it and I guess that's what made me dream about it.[9]

Cartwright surmised that Jerry's "general sense of incompetence was brought sharply into focus in the lab. This touched off similar memories from his past. Only one dream, the fourth, is complementary in theme and feeling to the others."[10]

Cartwright's and Jerry's observations align with emotional selection theory. While Jerry slept, emotional selection worked to alleviate his perceived incompetence, and then tested these changes with dreams. Throughout the night, his dreams progressed from targeting specific schemas that related to the here and now to more generalized ones. The good news for Jerry is his mind was actively working on reducing his perceived incompetence, though it will likely take months or years to rework such deeply rooted schemas.

While bizarre dreams typically occur during REM sleep, they occasionally occur during non-REM sleep. This suggests that dream tests can occur during emotional selection's accommodation phase. Likewise, while mundane dreams typically occur during non-REM sleep, they can also occur during REM sleep, suggesting that accommodations to schemas occasionally occur during the emotional selection phase. These observations agree with how systems are modified and tested in general. Engineers try to avoid making modifications to systems during testing, but occasionally they will fix defects in the field when errors are found so that testing can resume. For instance, during flight tests of the *Wright Flyer*, the Wright brothers found that their aircraft's wings did not generate enough lift because the camber (curvature) of the wings was too deep. Rather than bring their craft back to the shop for further modifications, they performed modifications in the field (literally). Likewise, emotional selection occasionally tweaks maladaptive schemas during dream tests, rather than waiting until the next non-REM sleep.

Taking Systems Offline

Imagine entering an office building and strolling toward the elevators, where you notice two technicians folding up out-of-order signs.

"Is the elevator okay to use?" you ask.

"We think so," the first technician says. "We just installed an experimental fall arrest system. We're pretty sure it will work."

"*Pretty sure*?" you ask incredulously. "Is it safe?"

"We don't know," the second technician responds. "It's untested. You'll be the first rider. Good luck!"

It looks like you are taking the stairs today.

Technicians and engineers installing untested parts in our everyday systems would certainly make life challenging. Sometimes the changes would improve our lives in some small way. Other times, they would vex or endanger us.

The scenario of technicians installing untested parts in an elevator is ludicrous because engineers and technicians routinely test systems offline before delivering them to consumers. NASA used a controlled environment for the FM-9 test discussed earlier. They went to great lengths to build a runway in the desert, install crash test dummies, remote controls, and several cameras on the airliner to perform and monitor the test. No one was endangered because engineers controlled the test environment.

According to emotional selection, sleep provides a controlled environment to safely modify and test our schemas. We no more want untested modifications to our schemas than we want an untested fall arrest system.

Describing sleep as a controlled environment helps explain why we require sleep at all. Drowsiness makes us feel as though we sleep to rest, but this intuition is unsupported by the evidence. Consider "healthy insomniacs" who require almost no sleep. Miss M was an elderly woman who slept mostly out of boredom.[11] Since childhood, she has slept for only about an hour per night and rarely experienced fatigue. During long bouts of wakefulness, she pursued her varied interests, including painting and writing. Rather than wanting to sleep, she often wondered why people bothered with it.

If we do not sleep to rest, then why do we sleep? One explanation is that we sleep to improve brain function. In 1994, neuroscientist Avi Karni conducted a simple experiment. He asked participants to focus on a location on a computer screen and then press buttons when either vertical or horizontal lines appeared in the background.[12] Participants returning the next day after a night's sleep were 15 percent faster at performing the task. The researchers deprived other participants of REM sleep by keeping them in the lab and waking them each time they entered REM sleep. The REM-deprived group showed no improvement in speed.

Since Karni's groundbreaking work, other researchers have shown that sleep improves performance. Harvard scientists asked participants to type a number sequence repeatedly on a keyboard.[13] After a few minutes, their performance typically plateaued. One group of participants returned later the same day, repeated the task, and showed no improvement. Another group that slept overnight tested 20 percent faster.

Abundant evidence suggests that sleep improves our ability to perform simple tasks, but its influence on complex problem solving is difficult to measure. An exception is a study by neuroscientist Jan Born. He asked a group of participants to solve math problems that involved a complex set of rules. Unbeknownst to the participants, the problems included a simple trick: the second number in the series of calculations equaled the previous problem's solution. Those who retook the exam after sleeping were more likely to catch the trick.[14] So, when faced with a difficult problem, the scientific evidence supports the notion to "sleep on it."[15]

Sleep also purges toxins from our brain. During sleep, neurons shrink to enlarge their spacing. The body then pumps fluids through the gaps, flushing out toxins accumulated during waking hours. Referred to as the *glymphatic* system, it is literally the "sanitary plumbing" of the brain. While this process also occurs during waking, scientists observed that it increases tenfold in mice during their sleep.[16]

While the fatigue that accompanies drowsiness may feel muscular, it is a matter of brain chemistry. The *basal forebrain* contains the cells responsible for maintaining wakefulness. The brain uses the neurotransmitter *adenosine* to shut these cells off. The longer we remain awake, the more adenosine accumulates in the basal forebrain, thereby causing drowsiness. So, although we feel "worn out" before bedtime, that tiredness we feel is literally in our head.

Sleep is important. Insufficient sleep leads to irritability, inattentiveness, dizziness, and even hallucinations. A study of medical interns working in intensive care units found that doctors on long, tiring schedules made 36 percent more serious medical errors and 460 percent more diagnostic errors than those with adequate sleep.[17] And exhausted medical professionals jeopardize more than just their patients. A nationwide survey showed tired interns were more than twice as likely to cause motor vehicle accidents driving home after an extended shift.[18]

Despite the importance of sleep, we often do not get enough of it. Adults average only 6.8 hours of sleep on weeknights instead of the recommended eight hours.[19] Many sleep only five hours per night during the week and compensate with nine hours nightly on weekends.

While erratic sleep schedules lead to mistakes, gradual sleep deprivation has been found to have virtually no effect on performance. In one study, researchers asked eight participants to reduce their sleep by 30 minutes every few weeks until each participant felt they did not want to reduce it further. Participants then maintained their shortest sleep duration for one month. The minimum duration achieved was 5.5 hours per night for two participants, 5.0 hours for four, and 4.5 hours for the remaining two. All participants showed an increase in sleep efficiency (less time to fall asleep and fewer awakenings). Researchers found no deficits in the many performance and psychological tests issued throughout the experiment. However, one year after returning to normal sleep schedules, all participants naturally slept less—between eight and 18 fewer hours per week. The evidence from this and other sleep-reduction studies suggests that most people can adapt to sleep schedules of about five and one-half hours per night.

Emotional selection describes the sleep cycle as providing the mind with an opportunity to modify and test its mental schemas. So, people who miss sleep also miss opportunities to adapt their schemas to waking life.

Emotional selection affects complex schemas, not the relatively simple schemas responsible for rote tasks, such as typing letters on a keyboard or driving. According to emotional selection, people who receive adequate sleep over months and years are, on average, more emotionally intelligent than those who do not. So, when you find yourself burning the candle at both ends, you are really burning opportunities to improve your mind.

Discussion

On average, people undergo 140,000 sleep cycles throughout their lives,[20] each of which provides the ideal environment for modifying and testing schemas. Would the brain and mind have ignored this opportunity to improve our schemas during our evolution? Unlikely. Instead, natural selection would have favored species that exploited sleep to improve schemas. In other words, natural selection would have selected emotional selection. (More on this in Chapter Six.)

Improving schemas while we sleep is clearly important, but how do these improvements translate into our waking lives? Quite simply, adapted schemas help us meet our needs. In the next chapter, we look deeper into the needs that occupy our dreams and the schemas responsible for meeting them.

THREE

Continuity of Needs

If you want peace of mind, do what you need instead of what you want.

— Santosh Kumar Pyasa

WE ALL HAVE NEEDS . Lots of them. Some needs, such as eating, drinking, and breathing, are physiologically essential. Other needs, such as those for belonging, love, and beauty, are less tangible. Our drives to satisfy our needs get us out of bed in the morning, motivate us to form relationships, and guide us when making decisions. We learn, work, and play to meet our needs.

Needs modify our behavior so profoundly that every branch of psychology studies human needs. Freud used the language of needs when describing his theory that the mind comprises three distinct processes: the id, ego, and superego. He described the id as harboring our most primitive drives, singularly focused on fulfilling basic needs such as food, safety, and sex. According to Freud, the id is oblivious to social norms and, unchecked, stops at nothing to satisfy its needs. The ego strives to meet higher needs, such as maintaining friendships and staying out of trouble with authorities. The superego strikes a balance between satisfying the lower needs of the id and the higher needs of the ego.

The psychological field of behaviorism studies needs by observing the behavior of people and animals. Ivan Pavlov experimented with the concept of classical conditioning by ringing a bell every time he fed a dog. During his experiment, he connected a tube to a dog's saliva gland and collected the saliva in a jar. Eventually, he found that ringing the bell without food present stimulated saliva production, suggesting the dog unconsciously associated the bell with food. Pavlov showed the physiological need for food could be associated with an unrelated stimulus. Behaviorists interpreted his study to mean that human behavior is conditioned unconsciously to satisfy the gamut of human needs.

Psychologist Henry Murray noticed our personalities play a role in how we prioritize our needs. In 1938, he published *Explorations in Personality*, which introduced a system of needs divided into *primary needs* and *secondary needs*. Primary needs are our basic, biological needs, such as food, air, and water. Secondary needs include achievement, nurturance, order, play, and love. He described how everyone shares these common needs and how personality determines the salience of the drives to meet them.

Murray categorized dozens of needs and described how some needs conflict. For instance, he described our need for autonomy as our drive to break free from constraints, and how this need can conflict with our need for affiliation, which he described as being close and loyal to another person. He also described how clusters of needs support other needs. So, his need categories "acquisition" and "construction" support the categories "achievement" and "recognition."

The psychologist most often associated with the study of needs is Abraham Maslow. In the 1940s, Maslow noticed that the needs of the monkeys he studied paralleled those of people. He also observed that monkeys fulfilled lower needs, such as thirst and hunger, before satisfying higher needs, such as friendship and love.

Maslow eventually turned his behavioral study techniques toward humans, noting that we, too, prioritize needs: "For the man who is extremely and dangerously hungry, no other interests exist but food. He dreams food, he remembers food, he thinks about food, he emotes only about food, he perceives only food, and he wants only food."[21]

Maslow defined eight categories of needs, with each category containing "sub-needs." For instance, his physiological category contains the sub-needs of food, warmth, sex, air, and other physiological needs:

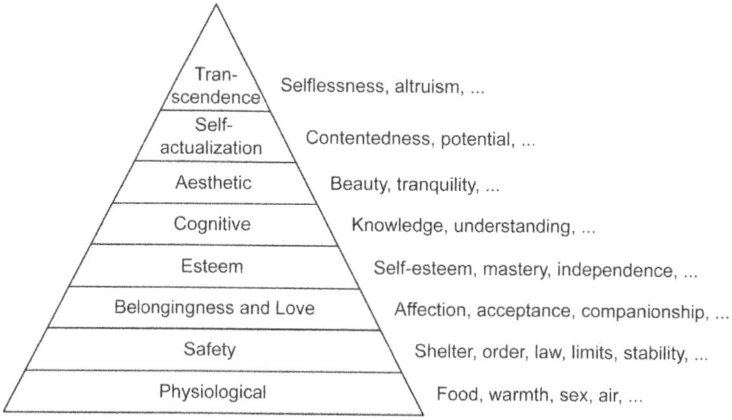

Maslow observed that a person's ability to meet basic needs profoundly affects his or her mental health. His clinical experience "consistently showed that individuals who satisfy their basic needs are healthier, happier, and more effective, while those whose needs are frustrated develop psychopathological symptoms."[22] Maslow also found that people whose needs were met as children were more successful in meeting their needs as adults.[23]

Rather than apply the traditional psychological study technique of observing the characteristics of people suffering from mental illness, Maslow preferred to observe the behavior of people he considered psychologically healthy. He paid particular attention to the steps they took toward meeting their needs.

The studies by Maslow, Murray, and other psychologists provide an excellent framework for categorizing the needs targeted by emotional selection during sleep. According to emotional selection, dreaming improves our ability to meet our full range of needs, from our basic needs, such as sustenance, sleep, and sex, to our higher needs, including belongingness, companionship, cognition, knowledge, and creativity.

Of the need categories described by Murray, Maslow, and others, I chose Maslow's in my peer-reviewed papers and in this book because of its simplicity and comprehensiveness. It is important to note that my use of Maslow's hierarchy is a mere preference. I encourage dreamers and researchers alike to consider which system of needs aligns with their experiences.

Unpleasant Dreams

Researchers often describe dreams as pleasant or unpleasant. Pleasant dreams contain beautiful or exciting themes, such as discovering pristine landscapes, finding exquisite jewels, or having sex. Unpleasant dreams contain sad or stressful themes, such as being rejected by a loved one, chased down an alley, or humiliated in front of a group of peers.

Unpleasant dreams test our ability to cope with unmet needs. For instance, while awake, we typically avoid dangerous places, such as precipices and dark alleys. When life thwarts our need for safety, we may be physically injured or suffer mental trauma. So, during sleep, emotional selection processes dreams to improve our ability to cope with thwarted safety needs and then tests these changes with unpleasant dreams of dangerous situations, such as falling or being attacked. Such dreams are universally terrifying, but are excellent tests of the schemas responsible for coping with the thwarting of our needs for safety.

While unpleasant dreams often include unmet physiological and safety needs, many incorporate higher needs. An 18-year-old student I interviewed dreamed:

> *My parents had to decide where to go for vacation. Our choices were either Cambodia or "Pandaland"* [a fictional theme park conjured in her dream]. *My dad at last decided that we would go to Cambodia. We arrived in our hotel in Cambodia, and the last thing I remember is that I went out for a walk and I passed a couple who was dressed in the same khaki shorts as I was.*

After describing her dream, the student professed her love for all things panda and elaborated that in the dream she wanted to go to Pandaland rather than Cambodia, as her father had proposed. Asked to deconstruct her dream, she said, "I do not make enough decisions on my own" and "both my parents are controlling." She expressed her frustration with her inability to sway her father's decision during the dream and felt that her dream tested her ability to assert her independence. While not discussed, the latter portion of her dream, in which she wore the same clothing as others, may have further evaluated her independence by testing how she felt about dressing identically to the group.

A classmate of the Pandaland dreamer was the captain of the high school hockey team. At the time of my interview, their season ended after being eliminated from a tournament. In the locker room afterward, several of his teammates cried, though he had not. That night he dreamed:

> *My teammates and I had one final practice at the rink after we lost the previous day in the tournament. I remember, at the end of the practice/scrimmage, we all gave each other hugs, especially the seniors. I remember I cried in the dream,*

> *and I thought it was particularly interesting, because I did not cry after our loss. Everyone else cried, too.*

The details of his dream mirrored his recent experience. However, his emotions changed. Emotional selection appears to have modified his schemas to allow him to be closer with his teammates, and then tested the modifications with a dream. The modifications appeared to have worked, as he grew closer with his teammates during his dream.

Pleasant Dreams

If unpleasant dreams test our schemas, what do dreams of flying, finding treasure, and other pleasant themes do? These dreams also test our schemas, but rather than testing our ability to cope with unsatisfied needs, pleasant dreams test how well we cope with satisfied needs.

The need to cope with adversity is obvious, but the need to cope with wealth, loving relationships, and other satisfied needs may sound like an oxymoron. Unfortunately, people with feelings of unworthiness and low self-esteem often show self-destructive behaviors. When faced with success, such people may sabotage their good fortune. So, after modifying schemas, emotional selection may process pleasant dreams to test whether newly modified schemas can cope with the happiness that accompanies our needs being met.

Consider flying dreams. With a simple push from the ground, dreamers take to the air, soaring above treetops and cities. A woman in her fifties dreamed:

> *Flying without an aircraft. Just soaring over the landscape at a high rate of speed, up and over trees, mountains, etc. Can only see through my own eyes what and where I'm going. Wake up exhausted but oddly at peace.* [24]

Gayle Delaney, a dream author who cataloged several common dream scenarios, noted that some dreamers struggle to embrace pleasant dreams of flying:

> Commonly, people dream of trying to fly and are unable to get off the ground. Or they may finally, after a lot of arm flapping and struggling, get into the air only to worry about losing altitude or about bumping into things or about getting tangled up with high tension wires. Others have trouble getting out of an enclosed space; they want to fly but can't get through the window or the ceiling and feel stuck. Sometimes the dreamer is trying to fly, trying to take off, but is unable to because someone is holding onto his or her foot and won't let the person go. Such dreams can be filled with frustration and anxiety.[25]

Delaney chronicled a pleasant dream in which the dreamer balanced anxiety and success:

> Frank dreamt that he had a chance to fly in the company of Winston Churchill. He was very excited, and he was trying to show Winston how able he was to complete complicated acrobatic feats in the air. He was having lots of fun and enjoying showing his prowess to Winston. Now and then he worried that he might not be able to keep his altitude, but he knew that if he worried for an instant, he would lose his altitude. So, he just kept his mind focused on the pleasure of showing off his skills.[26]

Emotional selection processes pleasant dreams to test our ability to cope when our needs are met. When dreamers experience increased stress during an otherwise pleasant dream, emotional selection may reject prior schema accommodations. Processing pleasant dreams successfully by embracing important dream moments signifies prior accommodations performed well and were likely adopted. So, basking in the comfort of a pleasant dream indicates that relevant schemas cope well with success while awake.

The Continuity Hypothesis

When reflecting on dreams, we often focus on their bizarre features, such as suddenly realizing that we can fly or finding ourselves back in our childhood home or school. However, much of our dream content reflects waking life. The values, fears, and desires we hold while awake are typically consistent with those in our dreams. People, places, and things in our dreams vary, but our personality remains consistent.

Dream researchers refer to the observation that some features of our waking life continue into our dreams as the *continuity hypothesis.*[27] Sleep and dream researcher William Domhoff summarizes:

> ... the content analysis of many dream series have led us to formulate what we call the continuity hypothesis. This hypothesis states that dreams are continuous with waking life; the world of dreaming and the world of waking are one. The dream world is neither discontinuous nor inverse in its relationship to the conscious world. We remain the same person, the same personality with the same characteristics, and the same basic beliefs and convictions whether awake or asleep. The wishes and fears that determine our actions and thoughts in everyday life also determine what we will dream about.

The wishes and fears that Domhoff describes can be framed in the context of needs. Wishes are the longing to fulfill our needs, and fears are the response to potentially thwarted needs. Using the terminology of the continuity hypothesis, emotional selection describes dreams as modifying and testing our schemas to satisfy our wishes and allay our fears.

The continuity hypothesis describes how only the emotionally salient features of our lives continue in our dreams. For example, while watching a horror film, we feel genuine terror—our heart feels like it is beating outside of our chest as the on-screen sadist approaches an empathetic protagonist. Despite experiencing these emotions while watching a film, studies have found they have scant representation in our dreams.[28] This observation aligns with emotional selection. Typically, only the emotional experiences important to our waking lives continue into our dreams, not the emotional experiences of fictional characters in a movie.

While the storylines of films and television rarely appear in our dreams, the needs we experience while awake routinely permeate our dreams. Study participants who went to sleep thirsty dreamed about their physiological need for water;[29] soldiers and other people who worked in dangerous environments often dreamed of their need for safety;[30] and one of our highest needs, our need for aesthetics, was found to frequent our dreams.[31]

While many studies have shown that individual needs appear in dreams, no single study tested for the full range of human needs. So, I designed a simple study. I had three judges assign Maslow's need categories to one hundred randomly selected dreams.[32] The judges scored each dream for Maslow's needs and whether the need was satisfied or thwarted. So, a dream in which the dreamer had enjoyable sex was scored as "satisfied physiological," while one in which the dreamer attempted to have sex but was repeatedly interrupted was scored as "thwarted physiological."

Dreams often address more than one need. A physiological dream of sex could include a loving partner, so judges could also score it as "satisfied belongingness." I compared the scores and assigned a "match" when at least two of the three judges agreed.

Many dreams contained categories in which all three judges agreed:

> *I am delivering a guest lecture for a colleague in a modern college auditorium which has a lot of high-technology audiovisual aids set up. The audience in the lecture hall is small, but a video camera broadcasts my face to a medium-sized television screen. I feel that I am not effectively making my point, which is about books written by a forgotten author. While still talking, I fumble around with cables, trying to connect the video signal to a larger screen. Finally, it works, and my image appears on a movie-sized screen hanging from the ceiling. I feel more confident now because this picture is much more impressive than before. I show one book by a favorite author—I can't remember who—holding it up to the camera for all to see. In it is a bibliography, which I read aloud. As I read each new book's title, I hold up that book, and its front jacket appears on the huge screen above. Every book is more obscure and harder to find than the previous one. These books are precious and meaningful to me, but I wonder whether I am getting through to the audience.*[33]

All three judges scored this dream as "satisfied esteem" because the dreamer expressed confidence and resolved the issue with the video connection while under pressure. The judges also labeled the dream as "thwarted esteem" because of the self-doubt expressed in conveying his enthusiasm to his audience.

Most dreams contained categories in which two of the three judges agreed:

> *I'm watching a film or tape taken from a helicopter, I suppose, of a city skyscraper all the way up to the roof where a singer—maybe an opera singer—is performing, singing to the sky in an evening dress. The "camera" passes over her and continues, showing the tops of other high buildings. Some other things happen. Now I'm in a car on a hill—it's night, been night all along—and I'm crying because I'm afraid to go up that high, I couldn't do what the singer was doing. The person I'm with hugs and consoles me.*[34]

The first judge scored this dream as "thwarted safety" because of the dreamer's overwhelming fear of heights, and as "satisfied belongingness" because of the support she received from the person she was with. The second judge matched the first judge for "satisfied belongingness," but scored "thwarted esteem" rather than "thwarted safety" because she focused on her irrational fears. The third judge scored "thwarted esteem" for the same reasons as the second judge.

Judges assigned thoughtful, plausible scores to the above dream yet arrived at varying opinions. My simple study shows the difficulty of deconstructing another person's dream. While dreaming, we use emotions and feelings to determine the significance of events. A stranger approaching could trigger fear if we believe he or she is dangerous, or arousal if we find the person attractive. Unfortunately, dreamers often omit such nuances when logging their dreams. Because the dreamer is the sole witness to his or her dream, the dreamer is often the ideal person to deconstruct it.

The judges matched one or more need categories to 92 percent of the dreams, confirming my hunch that dreams are chock full of needs. Next, I grouped the judges' matches into Maslow's basic needs (physiological and safety), social needs (belongingness and esteem), and personal fulfillment needs (cognitive, aesthetic, self-actualization, and transcendence).

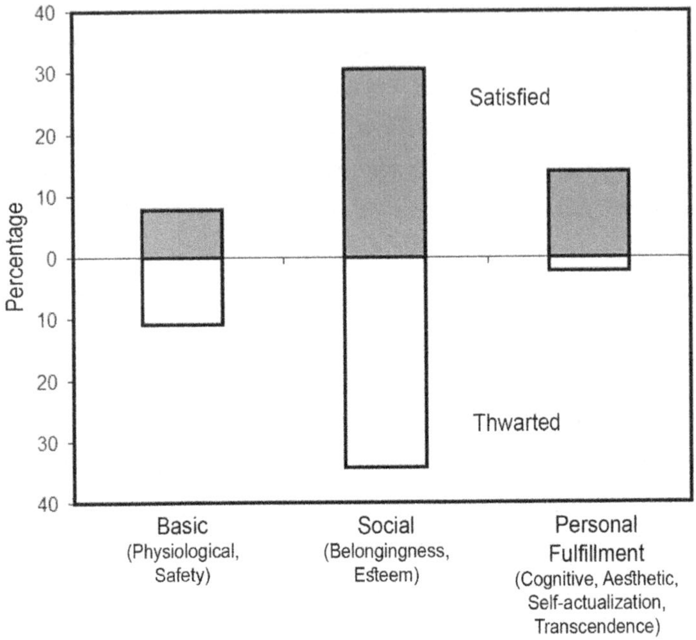

As shown in the above plot of the judges' scores, participants were more likely to see their lower, basic needs thwarted in their dreams. So, dreams of physiological and safety needs tend to be unpleasant, such as gasping for air while drowning or fearing for safety while being attacked. Dreams of middle, social needs, such as belongingness and esteem, tend to be mixed, such as finding or losing a loved one, or succeeding or failing at a task. Dreams of our highest personal fulfillment needs tend to be pleasant, such as finding yourself in a beautiful valley or fulfilling your cognitive needs by finding undiscovered rooms in your home.

The data of the above plot makes sense from an evolutionary perspective. Our earliest ancestors, for whom survival was particularly difficult, likely had a higher percentage of dreams that included lower, basic needs. Today, the social needs of belongingness and esteem dominate the scoring, which reflects modern society. So, while our day-to-day survival depends on satisfying our basic needs, our minds work on better ways to fulfill our higher needs while we sleep.

Discussion

Although I was excited to confirm that the gamut of human needs is well represented in our dreams, my study only demonstrated that we dream of needs. I had no findings supporting the idea that emotional selection helps to meet needs in the future. I needed more data.

In the next chapter, we will review my subsequent study, in which I surveyed over 15,000 people visiting an online dating website to determine whether their relationship needs continued into their dreams. I will then expand on my longitudinal study, summarized in the introduction, in which I tested whether participants who dreamed of a relationship saw their dreams come true.

FOUR

When Dreams Come True

Romeo: I dreamt a dream tonight.

Mercutio: And so did I.

Romeo: Well, what was yours?

Mercutio: That dreamers often lie.

Romeo: In bed asleep, while they do dream things true.

— William Shakespeare, *Romeo and Juliet*

After completing my study that categorized the human needs in dreams, I wanted to explore whether dreaming of needs predicted our ability to subsequently meet those needs.

The results of my prior study showed that people often dream of belongingness and esteem. These needs can be met during our waking lives through romantic relationships, as healthy relationships impart a sense of belonging, and experiencing love boosts our self-esteem. So, to test whether the needs we dreamed about were more likely to be met in the future, I decided to study a group of participants who were actively pursuing romantic relationships.

A Dating Survey

As mentioned, the online website Plenty of Fish generously hosted my study survey. Visitors to the site were greeted with a banner advertisement inviting them to take part in my study.[35]

The survey included an open question in which participants logged a recently remembered dream. Many people described their dreams of actual, real-life partners. A woman in her forties dreamed:

> *I dreamed about my ex. The feelings and emotion felt very real. I had dreams that he left me for another woman when we were married and they came true. And in that dream, I was trying to call him and couldn't find him. This last dream was of him looking for me.*

Rather than life partners, some participants dreamed of attractive non-partners, including co-workers, celebrities, and strangers. A single man in his sixties, who was not actively dating, dreamed of escaping his loneliness:

> *I dreamed that I'd met this beautiful lady. She was slim with long blonde hair. We went for a walk on the beach. She told me she loved me and was mine forever. I felt [like] the happiest man on earth. I don't know where we were. We held each other and kissed, then I awoke and found myself [alone] as usual. I laid on my bed alone and broke down and cried with loneliness.*

Many of the study participants reported dreams with sexual themes. So much so, that a woman in her forties expressed a mix of discomfort and humor relating hers:

> *Don't really want to describe it. Was quite sexual in nature and rather personal. Sorry! Jeff Hordley, the Emmerdale actor, was in it. (Please don't judge me!)*

There was no shortage of dreams with sexual content from both men and women. Here are some dreams shared by some of the male participants:

It was a sexually explicit dream about a threesome with a friend and my girlfriend.

I was out with my lover and we were at a wooded area, not sure where it was, but we had made love and were just enjoying nature.

It started out in a beautiful park near where I live. I was walking with a beautiful woman. I did not recognize her. We were holding hands and walking. We came to an outside stage and started making out. One thing led to another and we were making passionate love on the stage. When we were done, we noticed that a crowd had gathered in front of the stage, and they started clapping. That is when I woke up.

I was lying in bed and my roommate walked into my room and I was masturbating. I stopped and then he walked away. Then I started again, and he said from the hallway, "Is it really that important to you?"

I dreamed I was holding my future partner in my arms and we started making love all night.

I was a submissive slave being dominated by two women, forced to participate in demeaning sexual activities.

I don't want to go into detail but it was a very erotic dream about a close female friend of mine. Funny thing is, we have never done anything sexual together.

The above dreams share similar content and tone with the dreams of the female participants:

I was walking along a beach with a man. We were eating fish 'n chips, having a laugh and completely enjoying ourselves. [We had] sand in every nook and cranny, cold feet from the sea where we were jumping small waves, then slightly under the pier (mind the tide's out). We lay down, started kissing passionately, then it leads into passionate love making.

I was having sex with a married man while his wife was in the other room.

It was a sexual dream where I was trying to masturbate with a toy but I was being watched by someone. I stopped and put it up and proceeded to clean out the fridge.

I was sitting at a picnic table with my husband and some friends. Among these friends was a young woman I'd never met before. She had very short, very red hair, and she wore a turquoise men's-cut T-shirt with white/turquoise/green plaid shorts. She was very beautiful. I knew, somehow, that she was a bit gender-queer. Just my type. She sat next to me on the picnic bench, between me and my husband. Suddenly, she leaned over and laid in my arms. I held her for a few minutes, and then she knelt between my feet and kissed the tops of my breasts. We kissed, and I pulled her shirt up. I was on the point of asking her if it was okay to touch her breasts when I had to wake up.

It was my husband and I in a threesome. We were having the time of our lives. The orgasms were incredible. I woke up shaking hard from one.

I was up front at my used-to-be church, kneeling at the altar. Pastor comes over and lays his hand on my head. I'm looking down praying on my knees. I see his feet and they are bare. I start washing them like Mary did Jesus, then blot them dry, start rubbing and kissing them, then move up, unzip his pants, give him oral sex, then we have sex right there on the stage with everyone there. That's it. I know, I'm going to Hell, lol.

I was in a kitchen with a man who is a stranger to me, but in the dream, we were dating. We were just hugging and kissing and being affectionate with each other.

While both men and women reported lots of sex dreams, there were gender differences. Men were slightly more likely to dream of attraction overall, but the subject of their attraction differed. Rather than a past or current relationship partner, men were more likely to dream of attractive non-partners. Women, on the other hand, were more likely to dream of attraction to current or former partners. These findings agree with other studies that found men have higher drives for multiple partners while women have higher drives for monogamous relationships.[36]

My findings included lots of exceptions—many of the men's dreams contained romance, and many of the women's dreams contained random sexual encounters. Because we all have sexual and relationship needs, a healthy dream life ideally includes a combination of both types of these dreams.

Although my study collected lots of dreams that included attraction, most dreams had little or no bearing on relationships or sex. Work was a popular theme, as was being lost, being chased, taking an exam, and having technical difficulties. Participants with these dreams possibly dreamed about attraction on other nights, but the dreams collected for my study suggest their schemas were working on improving their abilities to meet other needs on this particular night.

Unsurprisingly, participants already in a relationship at the time of my survey were more likely to report being attracted to current or former partners in their dream, such as a past or present boyfriend, girlfriend, or spouse. Meanwhile, singles were more likely to dream about attractive non-partners, such as acquaintances, friends, or imaginary characters conjured in their dream.

Dating clearly influenced the dreams of participants. Singles who went on a date before sleeping were much more likely to dream of attraction. The dream of a male participant in his fifties vividly depicted the effect dating can have on dreaming. He reported that he recently broke up with a woman who would not commit to their relationship. Afterward, he dreamed of his lost love after a date with another woman:

> *I had a dream that we got back together and everything was great, then I woke up and felt terrible again.*

I also wanted to know whether having a crush on someone while awake influenced dream content. One of my survey questions read, "At the time of your dream, did you have a crush on, love, or otherwise have strong romantic feelings for anyone in your waking life?" Participants' answers showed that having a crush on someone profoundly affected dream content. Those who had a crush in their waking life were much more likely to dream of attraction, even if the crush was on someone participants never met, such as a celebrity or someone online.

Future Relationships

Emotional selection posits that people dreaming about relationships will be better at forming and keeping relationships in the future. But could this effect be measured? To investigate, I invited participants to receive a follow-up survey three months and twelve months later. These surveys collected participants' relationship status. Over one thousand people completed both follow-up surveys.

As mentioned in the introduction, dreaming of attraction was found to be associated with participants' futures. Singles who dreamed of a former relationship partner were 83 percent more likely to report being in a relationship in a follow-up survey. My study participants dreamed of a need for a relationship, and that need was met in the future.

On the surface, these results seem remarkable. Reviewing the contents of their dreams, however, we easily see the work performed by emotional selection. A woman in her forties dreamed of her deceased husband:

> *It was a dream where my late husband came up to me at a party and was angry that I was all alone, sitting in the corner. He wanted me to get out and meet people. I told him he was my husband, so I wasn't alone. He insisted I get out and meet men; he was so angry that I hadn't met anyone since his death. I was so confused because I kept telling him I wasn't alone, that he was my husband. He turned and walked away, and I grabbed his leg and wrapped my arms around his jeans and begged him not to leave me as he dragged me across the floor. I kept begging him to not leave me.*[24]

A woman in her fifties similarly dreamed:

> *I was somewhere in Hot Springs outside a bar and I kept hearing someone call my name. When I turned around and looked, it was my husband Ray who was killed back in 2002. I could see him and what he was wearing, and he was telling me something about it was OK to go on, to live my life, and he was smiling. I kept trying to tell him that I was OK, that I didn't need anyone else, that he was my soulmate, and I was OK alone. Then I heard this noise of some kind and I woke up. So, we never finished the conversation.*[24]

Schemas that were damaged by the loss of a source of love can limit our ability to meet our need for love. Emotional selection may target these schemas while we sleep, and then test them with dreams of lost loved ones.

Unrequited love is heartbreaking and can feel deeply humiliating, as reflected by the dreams of many participants:

> *I dreamt a few of my exes got together and talked about how stupid I was while I was sitting there with them.*
>
> *My ex-spouse having sex with her new love interest. She turned and smiled at me, then [her lover] smiled at me, then neither smiled at me. My children were there also.*

Several participants had pleasant dreams that tested their ability to sustain a future relationship. A man in his sixties dreamed:

> *In the first dream, [I] was traveling to [the first destination] but I wasn't alone. I had a new wife and we were very happy. Within a week's time, I had the second and third dreams that were a continuation of the first. This time our trip took us to [a second destination] then the third continued on to [a third destination]. The common point in these is the fact that I had a new and lovely wife.*[24]

Another man in his thirties was undergoing a divorce, yet he dreamed fondly of his spouse and kids:

> *I remember the feeling of comfort and [the] satisfaction of playing with my soon to be ex-wife and three-year-old son while seated at a picnic in a field where many other people were also picnicking.*[24]

Singles were not the only participants who benefited from dreaming of attraction. Participants who were already in a relationship at the start of my study were more likely to stay in their relationship if they dreamed of current or former partners, while those who dreamed of attractive non-partners, such as friends, acquaintances, or strangers, or who dreamed of unrelated themes, such as work or safety, were more likely to break up.

A woman who reported being in a relationship in both follow-up surveys dreamed of a future with her current partner:

> *Having a detached house in the country with my partner and future children milling about, playing.*

Rosalind Cartwright, a mentor of mine, published a study in 1991 that inspired my investigation into whether dreams can come true. During her tenure as director of Rush University's Sleep Disorder Service and Research Center, Cartwright wanted to know whether dreams helped people overcome the social trauma of divorce.[37] Seventy people volunteered for her study: forty who were depressed due to their dissolving marriages, and thirty who were free from depression.

Cartwright found participants with depression were more likely to have emotionally negative dreams and more likely to incorporate their problematic spouses as characters in their dreams:

> *My ex-husband told me to go to the hospital for a problem. It turned out I didn't have a problem. He was just giving me a hard time.*

> *I was having a fight with my husband. Yelling and angry. [He's] not taking good care of himself, overworking. He was running to me, not feeling well, [and] started collapsing. I was in a phone booth trying to make a call. I sat him down*

on a pile of something. I was persuading him to stay put while I called a cab to take him to the hospital. I was holding him up and said, "I have to go back and close the door."

Me and my husband and my sister went out to a restaurant like a pharmacy and sat in a booth and a co-worker was there who asked for lager if they would warm it. I went to do it and when I came back my husband said, "I'm divorcing you. Leaving you," and got up and walked out. I was stunned and embarrassed.

Those without depression either had milder dreams about their spouses or did not dream of them at all:

She and I had been out together. I dreamt of her closing a window when we arrived home. She wanted to close a window; the frame was old, made of small pieces of stone that fit together very precisely: Cotswold stone. It wasn't my home. It might have been her home.

I was invited to dinner by a friend and she said, "Sorry, Pat, (my ex) is not coming, so no dinner." So, I went home to bake a cake for myself.

Follow-up interviews one year later showed that participants who had emotional dreams about their spouses coped better. As a whole, their depression lifted and they took steps toward improving their lives, including dating, returning to school, changing jobs, and improving their health. Meanwhile, participants who had not dreamed about their spouses still struggled to adjust.

According to emotional selection theory, Cartwright's divorced participants who had emotional dreams reworked related schemas during sleep. Their emotionally charged dreams tested their new accommodations. If the accommodations appeared emotionally adaptive during their dream, they were retained, and the participants awakened better adapted to meet their future needs.

Our schemas define what it means to be a loving, supportive, sexual, and caring partner. The findings of both Cartwright's study and my study support the idea that dreaming of relationships helps us recover from past relationships and enter new ones. So, whether you have a terrible dream of an awful breakup or a blissful dream of love, the data suggests emotional selection is helping to ensure you are in a relationship in the future.

Discussion

This chapter whetted our appetite for the relationship between our dreams and our futures. In Part Two, I will delve deeply into how dreams prepare us to satisfy all our needs. So, people who frequently dream of work should be better performers at work; those who dream of safety should be better at protecting themselves and others; and so forth.

Why are dreams difficult to remember? Why do we sometimes know complete strangers in our dreams? Do animals dream? In the next chapter, I apply emotional selection theory to answer these questions and more.

FIVE

Frequently Asked Questions

Sleep is when all the unsorted stuff comes flying out as from a dustbin upset in a high wind.

— William G. Golding

IN THIS CHAPTER , I use emotional selection to answer common questions about dreams in a rapid-fire style. To avoid repetition, I will not revisit dream characteristics already covered in detail, such as why dreams can be bizarre and why dreams can be pleasant or unpleasant. Likewise, in Part Two, I will describe why we dream of common themes such as flying, falling, teeth and hair falling out, being chased or attacked, having trouble with technology, or failing to catch a bus or train. So, I will not discuss common dream themes here.

I briefly touched on a few topics earlier, such as recurring dreams and dreaming of strangers. We will explore these topics more thoroughly below.

Answers

At its core, a scientific theory must offer explanations. A good theory can also unify a range of seemingly unrelated observations under a single framework. For instance, Darwin's theory of evolution offers an explanation for the diversity of life on Earth under the framework of natural selection.

Now, let us put the theory of emotional selection to the test by reviewing how well it explains the phenomena of dreaming.

Why Are Dreams Difficult to Remember?

We dream for at least two hours during a typical eight-hour sleep, yet we rarely remember dreams upon waking. Why? Memory is a precious commodity that requires energy and other resources, so our mind dismisses memories it deems unimportant, including dreams.

As an analogy, consider the purchase of an automobile. Should the salesperson describe the thousands of modifications and tests performed during the automobile's development? No, because you only need assurance that the automobile is safe and reliable. Likewise, the mind is not concerned with remembering the dreams that modified and tested its schemas. So, instances when dreams are remembered upon waking are happy accidents.

As we will discuss in detail in Part Two, remembering and deconstructing dreams can lend insights into the schemas being modified and tested while you sleep. You can use these insights to further improve related schemas while awake. So, because emotional selection theory provides a tool to improve your schemas during wakefulness, there is now a benefit to remembering your dreams.

While we forget most of our dreams, there are measures you can take to improve your dream recall. Harvard researcher Robert Stickgold found repeating the affirmation "I will remember my dreams" four to five times before drifting off to sleep improves dream recall the following morning.[38]

Occasionally remembering your dreams does not mean they hold the same significance in your long-term memory as events you experience while awake. Ten years ago, I dreamed:

I was a guest on a talk show. There's a band rehearsing to go on. I realize the show is The Ed Sullivan Show *and that the band is the Beatles and that this is their inaugural performance. I noticed that the studio was very small, and there was only seating for fifty or so people, and that they make it seem like a bigger audience with camera angles.*

Though I remembered this dream upon waking, had I not happened upon it in my dream log, I would never have recalled having this dream. Had I truly been a guest on *The Ed Sullivan Show* and saw the Beatles live, the experience would be ingrained in my memory. Since the event occurred in my dreams, it was ephemeral. While a useful test for my mental schemas, the dream was apparently not worth the storage space in my long-term memory.

Do Children Dream?

While researchers typically believe that children as young as three dream, children's dreams differ somewhat from those of adults. For instance, the dreams of children tend to be short and often have violent themes.

With some prompting from a researcher, a four-year-old reported a dream:

His head came off. [Can you tell me anything else that happened?] *Nothing else. That's all.*[39]

The schemas of a child's mind are rapidly evolving, so emotional selection actively employs dreams to meet their needs. Because kids are vulnerable, they tend to dream of lower needs, such as physiological and safety needs. So, their dreams often contain scenarios of death, attack, or injury.[40]

Does Everyone Dream?

All healthy brains dream. Some people believe they rarely or never dream, but laboratory experiments have shown that even these people typically report a dream when woken from REM sleep.

Tricyclic antidepressants suppress REM sleep. When the brain loses the ability for REM sleep, it cannot process REM dreams. Because these dreams test accommodations, losing REM sleep would impair emotional selection's ability to experiment with its schemas, reducing its ability to improve them. When a person stops taking these antidepressants, REM sleep dominates the later sleep cycles, as the brain apparently "catches up" on REM sleep. Emotional selection explains this observation as schemas that are deprived of REM sleep create a "queue" of dream tests to be processed. After REM sleep is restored, emotional selection processes the queue, resulting in a flurry of dreams.

Why Do I Dream of Strangers?

In our dreams, strangers appear in various forms: thugs chasing us down dark alleys, travelers watching us catch a bus, and attractive lovers with whom we share passionate sex. Our minds retain countless memories of people we know, so why do we dream of strangers?

Because our schemas must cope with the familiar and the unfamiliar. While navigating our lives, we must interact with those close to us and those new to us. So, we dream of strangers simply to help us cope with strangers.

Consider a woman recovering from a breakup who met an attractive stranger in her dream:

> *Dreaming about a face that I did not know, a male, a new friend. He was handsome and smart and really liked me.*[24]

Meeting a new special someone is often the first step toward entering a romantic relationship. Our schemas must prepare for such future encounters. By dreaming of an attractive stranger, the woman's schemas prepared to meet an attractive stranger in her future. Strangers take many forms in our dreams. As explored in the next question, we can even conjure strangers we know in our dreams.

Why Do I Sometimes Know Strangers in My Dreams?

Dream scenarios that incorporate important, albeit non-existent, characters in our lives showcase the remarkable ability of the human mind to create an alternate reality in our dreams.

A single woman in her twenties dreamed:

> *I was back in high school, out of nowhere I see this really cute guy, seemed like I knew him like forever, just comes up to me, smiles and kisses me. I felt happiness, we walked around the school, talked and kissed more, then cuddled during [physical education class]. The background disappeared into a house, I'm guessing it was his house ... he disappeared, and another guy walked in, said it was his brother who was also trying to make a move on me, [my boyfriend] walked in and started arguing with him. He gets my hand and leads me straight to his room ... looked at me and whispered a few words that I didn't seem to understand.*[24]

The dreamer expressed that it "seemed like I knew him forever." Her familiarity with the fictional boyfriend in her dream activated different schemas compared to an attraction to a complete stranger. Knowing the characters in her dream also allowed for more complex dream sequences, such as the rude brother, than if the dream characters had been mere acquaintances.

As a woman in her twenties, she likely has ex-boyfriends in her waking life who could have stood in for her conjured boyfriend. However, such real-life characters in our dreams can bring unintended emotional baggage, such as memories of a contentious breakup. So, her mind conjured fictional dream characters that were baggage-free.

Dreams prepare our schemas to meet future needs. A dream featuring fictional strangers that seem familiar to us, whether they appear as boyfriends, family, friends, or bosses, helps prepare our schemas for possible future encounters. Remarkably, our mind can convince us that we know a person intimately in our dreams, even if it conjured that person only moments ago.

Why Do I Dream of Celebrities and Other Famous People?

Dreams often incorporate famous people, including historical figures and celebrities. A woman in her thirties dreamed of hanging with famous friends:

> *It was just a gathering, kind of seemed like a party, but nothing wild. Famous people there, and it was as if I were really great friends with them. Just talking, taking pictures, and remembering things from the past.*

A woman in her forties dreamed of touring with a pop star:

> *I was on a tour bus with a famous pop singer. We were out west and stopped at a small dusty town. I got out with the band, feeling famous, too. Went to the store and the bathroom. There was no toilet paper, but a man heard me and handed it to me through the curtain of the door. Went back to the bus feeling great [with] two girls screaming, "It's Justin Bieber!" I was walking with him.*

Meeting famous people can be exhilarating and intimidating. Their positive attention may inflate your self-esteem, while negative attention may have the opposite effect.

To appreciate the cameos famous people make in our dreams, consider how the interaction made you feel during the dream and how the exchange would feel with a less famous character. Right or wrong, society elevates the importance of the famous, and our schemas characterize this understanding. When meeting someone new, we expect a certain level of anxiety and excitement. These feelings elevate when interacting with a famous person. This explains why emotional selection conjures famous people in our dreams—to intensify a dream test.

Why Do I Dream of Family Members Who Do Not Exist?

People often conjure family members in their dreams who do not exist in their waking life. A woman in her twenties dreamed:

> *A glimpse-into-the-future type dream. I was living in Washington, I had a one-year-old daughter, I was just finishing school and working as a high school substitute teacher (I'm going to school to be a teacher). My baby's father took off and just left me with the baby, barely able to support myself. An old boyfriend whom I am good friends with now flew up to see me and help out for a while. During his stay with me, we rekindled what we had and started a relationship.*[24]

People wishing for or expecting children must prepare. Every parent knows firsthand the life changes that arise with starting a family. To prepare for a future that includes children and a spouse, schemas may accommodate changes and then test these changes with fictional characters that represent our potential family in the future.

While the above dream of fictional family members included an absentee father, dreams of a future family are often heartwarming. A woman in her thirties dreamed of future children:

> *Having a detached house in the country with my partner and future children milling about, playing.*[24]

Finding yourself in a family of strangers would be truly bizarre. To prepare us for a future family, emotional selection often conjures fictional characters known to us in our dreams, including family members.

Why Am I Attracted to (Insert Name Here) in My Dreams? He/She Is Repulsive!

Occasionally, you may find yourself attracted to someone in your dream who you find unattractive while awake. Consider this dream of a woman in her forties:

> *I dreamt I was having an affair with my boss to whom I am not remotely attracted and my mother found out who I am estranged from, and she was furious. All the time things were falling down and trapping people whom I knew but can't remember, nor [could I] see their faces, and I couldn't help them.*[24]

During dreams, we occasionally do or say things we would never consider doing or saying when awake. However, this does not necessarily mean we harbor secret desires. Instead, emotional selection places us in these situations to test our schemas. The dream in which the woman had an affair that her disapproving mother discovered was likely testing schemas related to the mother, rather than schemas related to her boss. The affair simply gave her mother something to judge. She could alternatively have dreamed of embezzling money, assaulting an innocent person, or another scandalous act.

Deconstructing dreams in which you do things outside of your values can be challenging. It may be tempting to conclude that a dream of having an affair with a repulsive person tested schemas related to your desire to enter a similar relationship while awake. If the dream truly feels out of character, review its theme for other, more subtle themes. It is likely that your dream tested schemas tangential to your dream's central theme.

Why Do I Dream of Upcoming Events in My Life?

Do you have a big presentation this week? Or a date with a certain someone? Upcoming events often permeate our dreams.

A woman in her thirties dreamed:

> *I was at an upcoming convention. It was the day before, and I was helping with support staff duties. The organizer asked the head of guest relations if anything was needed for the "party" room; the head of guest relations indicated that she would be taking everyone out that night and notified me that I was a part of that group. A short while later, a member of the hotel staff came to find me to let me know that there was a problem with my room and it wasn't ready yet.*[24]

Dreams help us prepare for a potential future, whether that future is days, months, or years away. This preparation runs the gamut from long-term goals, such as finding a life partner or pursuing a career, to singular events, such as presenting before influential people. Regardless of your goals, if future events occupy your waking thoughts, there is a good chance they will occupy your dreams.

Can Dreams Predict the Future?

Dreams are not prescient, like the fabled crystal ball. Consider the common dream of finding riches:

> *I was winning the Publisher's Clearing House sweepstakes and spending the money on things that I need, such as a brand-new car and a house. Buying new clothes for my son and me and traveling to visit friends and family. I specifically remember that the SUV I bought had 40,000 miles on it.*[24]

And the common dream of dreadful accidents:

> *In my dream, my best friend told me that her father died, and then my father walked in the room and told me that one of his friend's wife died. I didn't know whom to console, and I was overcome with sadness and crying, then I woke up.*[24]

The above dreams would neither increase the dreamers' chances of winning the lottery nor of losing a loved one. However, the dreams would improve the dreamers' ability to cope with newfound wealth and grief.

As we saw with my dating survey, participants who dreamed of attraction were more likely to meet their need for a relationship in the future. According to emotional selection, these participants improved their ability to meet this need through dreaming. Dreams are neither mystical nor magical. Instead, dreams aid in evolving the schemas that mold your future.

Why Do I Dream of Zombies, Giant Spiders, and Other Monsters?

Our dreams often incorporate fantastic scenarios. A woman in her twenties dreamed:

> *The world was ending because of an epidemic and people were becoming zombies, like everyone who wasn't infected was trying to find a safe haven. I kept running away from these people and found myself getting into a van with a bunch of children and people. I was the driver and had to start the van by hot wiring it. I had difficulty doing [it], and a young [man] appeared and showed me that you had to take the red and green wires and you had to intertwine them together. I drove off onto the street in search for a new safe haven for these people and myself.*[24]

A man in his thirties dreamed:

> *At a mini-mall, kind of underground—a very self-contained mini city type mall—I was driving a car around, trying to round up the last living American presidents. I managed to get them all, and just as we crashed into what we thought was an exit, [it] turned out to be some kind of zombie lab. Zombies (lab professor zombies) were breeding other zombies. George W. II sacrificed himself so that the rest of us could escape. We barricaded ourselves in the only empty room. Then Obama turned. He was a full-blown zombie within seconds. I woke just as he went for my head.*[24]

We do not need to defend ourselves from zombies and other fictional threats during our waking life, so why dream about them? Because fictional threats provide extreme tests of our schemas. If a flesh-eating zombie closed in on you from one end of a corridor and the neighborhood bully from the other end, in which direction would you run? Toward the insecure bully, of course.

Anything is possible in a dream. If your mind decides it needs an extreme test of your schemas, then it conjures extreme themes such as flying, falling, or being chased by the undead.

Why Do Dream Themes Recur?

A single sleep cycle sees small changes to our schemas. Significant improvements accumulate over weeks, months, or even years. When emotional selection modifies and tests the same schemas repeatedly, we may find the same cast of dream characters repeating the same actions. Other times, dream themes recur, but the characters and circumstances change.

As a child, I was outgoing with friends and family, but shy in group settings, such as school and organized athletics. During this time in my life, I often dreamed of humiliating situations, such as finding myself accidentally naked in public. For the reader unfamiliar with this dream, its theme can be dreadful. I still remember one such dream I had as a teenager:

> *I was back in grade school, queuing up for the water fountain, when it occurred to me that I had forgotten to get dressed that morning. Initially, I was the only one aware of the situation. One-by-one my classmates noticed my predicament. I was mortified.*

As I matured into adulthood, the dream theme followed me, though the scenarios varied. Sometimes, I was an adult in the dream, having forgotten to wear my clothes to work. Other times, I was back in grade school, on the playground, or in the classroom.

Gradually, my reaction to my dreams of being accidentally naked abated. Rather than feeling humiliated, I developed a more self-forgiving attitude. "Oh, well, everyone makes mistakes," I would say to myself during the dream, embarrassed but not devastated. This trend continued until years ago, when I had the dream yet again. The theme was the same, but this time I laughed at myself during the dream when I realized how ridiculous I looked. I have not had the dream since.

My ability to cope during these dreams tracked my waking perceptions. As I matured into adulthood, the self-consciousness I experienced as a child faded, as I realized everyone made mistakes and that humiliation was not fatal. I now understand that, as I slept, emotional selection modified the schemas that made me overly self-conscious. Because my schemas were deeply rooted, the modification and testing took years. As my anxieties subsided, so did my corresponding dreams.

Why Do I Relive Past Trauma in My Dreams?

Trauma comes in varying degrees and forms and includes emotional and physical trauma. Soldiers who survive a battle often relive the experience in their dreams. A Vietnam veteran dreamed:

> *I was right back there, just the way it was. Shells were bursting all over the place. My buddy was hit by a shell right next to me. There was blood and screaming. It was just the way it was. Please don't ask me any more details.*[41]

After emotional selection repairs the damaged schemas of trauma victims, replaying the traumatic experience tests the new accommodations. Unfortunately, it often requires the dreamer to relive the trauma.

Because emotional selection repeatedly modifies and tests damaged schemas, the dreams of trauma victims often recur. As schemas improve, the recurring dream incorporates additional schemas, resulting in changes to the content of the dream. This broadening of content often results in a dream progression from realistic themes to increasingly abstract and bizarre ones.[42]

A war veteran turned to the psychologist Carl Jung for relief from his wartime nightmares. Note how his dreams progressed from realistic to fanciful:

> *In the dream, the man is in his home and suddenly becomes terrified. It is night. He goes to the front door and locks it. Then the back door. He locks all the windows on the first floor. But the sense of terror and panic continues to build, and he goes upstairs and locks all the windows, but just as he begins to close the last window a grenade explodes outside the window. This dream recurs again and again during three months of analysis, until suddenly one night, when he goes to close the last window, a roaring lion appears and the dreamer wakes in terror.*[43]

The soldier's dreams enacted a realistic wartime threat. His recurring dream theme initially showed no progression, suggesting his damaged schemas failed to improve. Eventually, his schemas appeared to have adapted, as the last dream replaced the realistic grenade with the fanciful threat of a lion. His dream progression suggests that his schemas satisfactorily handled the specific threat of a grenade and moved on to more abstract ones. While the dream of the lion terrified the man enough to wake him, the dream progression into abstraction signified that his schemas improved.

The dreams of people with post-traumatic stress disorder often lack a dream progression to abstraction, as their recurring nightmares remain realistic.[44] This observation suggests that emotional selection is failing to heal their damaged schemas. Because of this lack of progress, their nightmares appear stuck in an unsuccessful loop, playing the same dream repeatedly as their minds produce accommodations that emotional selection continually rejects.

Why Do I Dream of Fictional Trauma?

Even people fortunate enough to have avoided physical and emotional trauma must protect themselves. Disturbing stories of assault and murder fill the news, books, television, and movies. So, our minds may prepare for future trauma by dreaming of fictional trauma to modify and test related schemas.

A woman in her fifties dreamed:

> *A man kidnapped me and made me his sex slave. He beat me a lot. I tried to run away, but he always found me. He had also kidnapped a child, but I helped him/her escape.*[24]

Despite the horrific nature of her dream, her fortitude helped her attempt to escape and even tend to a child in need. Such dreams test changes to schemas intended to help cope with traumatic events. By developing coping skills for trauma during our dreams, we improve our ability to weather horrific events while awake.

Why Do I Dream of Committing Crimes?

While most dreams with themes of violence cast the dreamer as the victim, dreamers occasionally dream of being a perpetrator. Often, such dreams are crimes of passion.

A woman in her fifties dreamed:

> *I dreamed that my boyfriend and I were at this house and we were all together—him, his ex-girlfriend, and me, and a few minutes [later], both got up and left, and when I went to find them, they were in the bathroom, and he was standing naked, and she was on the commode naked and crying and said, "I can't have sex with you if we're not together." And so, I stabbed my boyfriend with a knife in the neck.*[24]

A man in his twenties dreamed of murder and suicide:

> *I killed my ex-girlfriend and her ex-boyfriend then went to kill myself and woke up.*[24]

A woman in her sixties dreamed of covering up a murder:

> *Dreamt I murdered someone and hid the body in a cabin, then years later when cleaning the cabin, I remembered the murder. But with many of the people I know working at it, I thought it best not to let them know about the body, so I took steps to help them avoid it. They found it anyway, and I was feeling a little sorry that they all looked at me like they were mad. I felt nothing whatsoever about that, except curiosity, and no connection to the potential liability of arrest. Kind of disturbing. Had to think when I awoke if I had really done that.*[24]

Committing crimes in our dreams does not mean we intend to do likewise while awake. However, such dreams do demonstrate that we need to cope with ethical challenges in our lives. As with all need categories, dreams that test our needs associated with ethics range from mundane to extreme. Murder is the ultimate form of aggression, so it serves as an extreme test of our schemas' ability to cope with poor ethical choices. However, as with all dreams of extremes, dreams of murder do not suggest we are preparing for murder any more than flying dreams prepare us for being weightless.

Why Do I Dream of Being a Kid Again?

While dreaming, adults often find themselves transported back to their childhood.

A woman in her forties dreamed of her siblings:

> *I was a kid again and was learning my brother and sister were being given special treatment by my parents on occasion after occasion and I was completely forgotten. At times they didn't even remember I existed. I was frustrated and sad and angry, but helpless to do anything about it.*[24]

Childhood is an emotional period in our lives. Consequently, dreams of childhood provide an excellent backdrop for extreme tests of our schemas.

A man in his fifties dreamed of being comforted as a child:

> *I dreamed about my dead father and my mum. They were in a room, and I was young and seeking comfort from them over something that had upset me. When I gained entry to the room, it was a bedroom of a house we lived in when I was about two to ten years old in England. I could almost touch the wallpaper and smell the smells of the old house, it was so realistic.*[24]

The man's dream returned him to his childhood home, where his parents provided the love and security he once knew. Emotional selection created his dream to test schemas related to his need for comfort from important people in his life.

Why Are Some Dream Characters Faceless?

Encountering a person with no mouth, nose, or eyes would be a remarkable experience in reality. However, such encounters are commonplace in our dreams.

A man in his twenties dreamed of faceless parents:

> *In my room, it felt real enough till I walked downstairs and I went to talk to my parents. … When they looked at me, they had no faces, it was just a mass of skin where their faces should have been.*[24]

A single mom in her thirties dreamed of a passionate encounter with a faceless man:

> *I was passionately entangled with a faceless man. My heart and my emotions were running wild, and I knew from the depths of my soul that I was in love. I was safe and content. I knew he wouldn't hurt me. I knew he was my mate. I was happy and satisfied with my life. Suddenly, we were all at the park together, and my ten-year-old son was playing ball with the faceless man. I could only see the man's back as he caught the ball. My son was so happy. He told me that he really liked this man and he wanted to pick him to be his daddy. Then we were at home, all of [a] sudden. He came up behind me while I was in the kitchen and kissed the back of my neck and hugged me and told me how much he loved me and how happy he was.*[24]

Dreams often omit irrelevant details when modifying and testing schemas. The latter dream included the woman's fictional soulmate. She loved him during the dream. The man's facial features were inconsequential compared to his more important characteristics, including being a father figure for her son and being a loving partner. His face was literally immaterial, so her dream simply omitted it.

Why Do I Dream of Scenes From TV and Movies?

As mentioned, watching a film rarely affects our dreams, but, like most things with dreaming, there are exceptions. Through transference, you may share the emotions and feelings experienced by characters in movies and television programs. During sleep, dreams may modify and test related schemas, making excerpts from movies and programs excellent testing fodder. However, during the dream, we are often no longer bystanders empathizing with the characters; we are the characters, experiencing the emotions firsthand.

Guy had a dream after watching *Honey, I Shrunk the Kids*:

> *The dream begins in my lab. A very important woman doctor is inspecting us. She comes and looks over my shoulder as I pour some concoction with formaldehyde and other ingredients into a test tube. I explain that it will help me observe some forms of life without actually harming them. She says, "Let's see if it works." She pours some of the liquid onto the top of my head and, the next thing I know, I am in the test tube, swimming around in this thick cloudy stuff. I have shrunk! This doctor is turning the tube round and round, examining me. I feel dizzy and manage to get my head up in the air and yell, "Hey! Let me out! This is awful!" So, she shakes the tube and I fall out with a thump, feeling glad to be back to normal.*[45]

Guy confessed to having control issues and being a sexist who felt threatened by women gaining power within his corporation. Watching the onscreen predicament of the shrunken characters apparently triggered his fears of losing control in his own life. During sleep, emotional selection attempted to address his maladaptive schemas. The theme of the movie, which triggered the schemas, provided the perfect dream test. Apparently, his mind decided to turn things up a notch by placing him at the mercy of a female colleague.

Why Am I Sometimes an Observer of My Dream?

Whether we are running from zombies in our dreams or making love with our crush, we typically experience dreams from a first-person perspective. However, occasionally we are merely observers of our dream. These dreams can feel like out-of-body experiences as we watch ourselves interact with others. Alternatively, we may not be a character in the dream at all. Referred to as "vicarious dreams,"[46] we observe the events of these dreams as if watching a movie or play.

A woman in her forties dreamed of observing another woman:

> *It was a woman running with her kids. It was like a movie and everywhere she went, her husband found her and the kids. She tried to escape but couldn't and one day she went to the cops, and [they] told her they couldn't do anything until he did something. Then she got away after [a] long while and, somehow, he tried to kill the cop, and he shot him.*[24]

A man in his twenties dreamed of watching himself:

> *My friend and I were on a date, and a tornado came. He got called to work, and all I really remember is standing like an out-of-body experience watching everything happen, watching the destruction and watching him risk his life to save and rescue everyone else.*

Empathy is the ability to understand and share the feelings of others. To have nurturing, supportive relationships, we must empathize. This need to empathize also applies to the self. Many of us are more forgiving of others than we are of ourselves. Out-of-body dreams literally give us a new perspective. Much as we sympathize with other characters in our dream, observing ourselves may work toward improving our ability to meet our needs for self-sympathy and self-compassion.

Why Are Some Dreams Lucid?

Dreamers typically perceive dream scenarios as actual events. Lucid dreaming is when dreamers become aware they are dreaming. As explained by Robert Stickgold:[47]

> Lucid dreaming, which involves being aware of the fact that you are dreaming while you dream, appears to be a state between REM sleep and waking. Regions of the

> prefrontal cortex, which control logical reasoning and executive decision-making and are normally turned off during REM sleep, appear to be turned back on. This allows some lucid dreamers to gain partial control over the events in their dreams.[48]

Consider a lucid dream from my dream log:

> *Bullies were hanging around my car. As I approached, I realized they were going to beat me up. My dream then became lucid and I had the power to toss them around with an invisible force of concentrated blows that I could emit from my hands by concentrating. My powers were not fully under my control. I needed to concentrate and have confidence to execute them. Even then, they did not always reach their full potential. At one point, I used them to levitate, sort of like Iron Man, though I had mixed success.*

Despite knowing I was dreaming, my dream was still emotional. I felt empowered when giving the bullies their comeuppance. However, the dream also felt like a good test of the schemas that affect my sense of competence, as I had mixed success wielding my newfound power.

Many people cultivate lucid dreaming by taking steps to improve their ability to recognize when they are dreaming. Proponents have made claims about the powers of lucid dreaming, saying it accelerates healing, expands creative problem solving, and improves memory management. However, most lucid dreamers simply use their ability to change the course of an unpleasant dream:

> *I find myself at the edge of a frightful precipice, the mere sight of which makes me tremble: a sheer, or even overhanging cliff many hundreds [of] feet high. At the bottom are sometimes*

> *sharp rocks, sometimes houses and trees which look small in the distance. At the moment when I tremble and hold tight, the dream suddenly becomes conscious: I realize that I am dreaming, that all this is illusory and that I am in no real danger. Then, in order to see what will be the result of this decision, I make up my mind to throw myself into the abyss. I do so and I always arrive at the bottom without a shock, unless my fall ends in a delightful flight.*[49]

Rather than altering a dream, some dreamers use lucid dreaming to terminate an unpleasant dream, as demonstrated by a man who survived a bombing in World War II:

> *Our house was hit by firebombs, and all the houses around were blazing. Heavy bombs were coming down thick and fast, and although I had been very frightened, I was running for water when this feeling happened. I said to myself, "This is the end. We are all going to be killed. There is no escape. We are a sitting target." [Realizing it's only a dream], I became detached and very calm, and my fear vanished. It was all unreal, and I wondered when I was going to wake up.*[50]

REM dreams test our mental schemas, so taking control of a dream effectively cancels the dream test. Any such cancellation, whether because of lucidity or being woken by your morning alarm clock, will impair the ability of emotional selection to adapt your schemas.

The firebombing dream tested schemas similar to those of the soldier's grenade dream we reviewed earlier (see Why Do I Relive Past Trauma in My Dreams?, page 71). For these dreams to be effective tests, they must trigger fear. The firebombing dreamer became detached and calm when his dream became lucid, which canceled his dream test. Lucid dreaming may spare us from nightmares, but nightmares are dream tests, too.

As we will explore in Part Two, you can take steps while awake to improve your ability to influence dream content without evoking lucid dreaming. Therefore, I recommend people work while awake to improve the schemas tested by their dreams, rather than using lucid dreaming to cancel dream tests.

Do I Solve Problems in My Dreams?

In the 19th century, chemist Friedrich Kekule was at an impasse while trying to derive the chemical structure of benzene. After several unsuccessful attempts at modeling it as a string, he dreamed:

> *I turned my chair to the fire, after having worked on the problem for some time, and dozed. Again, the atoms were gamboling before my eyes. This time the smaller groups kept modestly to the background. My mental eye, rendered more acute by repeated visions of this kind, could not distinguish larger structures, of manifold conformation; long rows, sometimes more closely fitted together; all twining and twisting in snakelike motion. But look! What was that? One of the snakes had seized hold of its own tail, and the form whirled mockingly before my eyes. As if by a flash of lighting I awoke.*[51]

Upon waking, Kekule successfully worked out the mathematics of benzene as a ring rather than a string. While a fantastic discovery, the idea that his unconscious mind solved his problem during his dream is unlikely. The dorsolateral prefrontal cortex—the part of the brain used for complex problem solving—is disabled during both non-REM and REM sleep. Furthermore, he describes how he worked out the complex mathematics upon waking, not during the dream.

During a dream, we often resolve straightforward dilemmas, such as eluding an attacker or finding a treasure, but the capacity to solve complex problems is lost to us. Though I studied advanced mathematics, I cannot do basic arithmetic during dreams. In one, a man told me his birth year. Doing the simple arithmetic, I calculated his age. In the morning, when my dorsolateral prefrontal cortex and I woke up, I realized I was off by eight years.

It is more likely that Kekule's dream of whirling atoms inspired his solution upon waking, rather than a dream solved his problem during sleep.

Can Dreams Be Interpreted?

As mentioned, dreams are neither symbolic nor metaphoric, so they require no deciphering or interpretation. This description of dreaming aligns with Harvard dream researcher Allan Hobson's opinion on the literalness of dreaming. Hobson cited his dream of a renovation calamity caused by a sloppy neighbor and concluded that the meaning of dreams is in their emotional salience, not in hidden meanings.

Here is an excerpt from Hobson's dream:

> *Then we are in a house, not at all like mine but assumed by my dreaming brain to be mine, and Richard's friend is spray painting the white wall (we have none in our house) with blue paint (neither do we have any blue rooms). The paint sprayer is a tank device of the type used to apply copper sulphate to grapevines or to exterminate cockroaches. Suddenly, the paint is being sprayed not only on the wall, but upon a painting hanging on the wall.*

> *My fears are confirmed. I yell at Richard to bid his friend stop.*
>
> *For some reason, he has to go upstairs to turn off the machine (although it appears to be fully portable and self-contained), and this takes an inordinate length of time as the painting continues to suffer.*

Hobson explained his dream:

> For me, the meaning of the dream is transparent: I am anxious about my property and about entrusting it to people who are careless about their own houses. This characteristic, known in psychological terms as emotional salience (or relevance), is all I need to understand the dream … I see no need and no justification for treating this dream as a disguised, symbolic expression of anxiety about other related themes.

Dreams have no hidden meaning. We will explore this idea in detail in Part Two, in which I deconstruct several dream themes and discuss their relationship to the needs we strive to meet while we are awake.

Do Animals Dream?

Both mammals and birds undergo REM sleep. Dog owners often observe that the legs of their sleeping pets mimic running during, what appear to be, imaginary chases. My childhood dog Cocoa even yelped during her dreams—the human equivalent of talking in her sleep. Occasionally, she barked loudly enough to wake herself. She then looked around, apparently concluded she was dreaming, and dozed back to sleep.

Research on the sleeping brains of cats provides intriguing evidence that our feline companions also dream. French researchers removed a cat's pons—the portion of the brain that paralyzes muscles during sleep. They found that when these cats entered REM sleep, they leaped to their feet and presumably acted out their dreams, which included chasing invisible prey and hissing at invisible adversaries.[52]

Perhaps the most compelling evidence comes from our closest relatives, the primates. Koko was a lowland gorilla who was taught sign language by her caretakers. One day, she awoke distressed. When the researcher signed, "What's the matter?" she signed, "Dark, bad, dirty, awful." When the researcher inquired further, she signed, "I don't want to talk about it."[53] Such interactions have led her trainer and lead researcher, Francine Patterson, to conclude that Koko dreams.[54]

Given that every animal's mind contains schemas, the content of animal dreams is likely similar to our own, featuring scenarios of being chased, attacked, and having sex. They likely also have pleasant dreams. It would be nice to think that pigs, too, can fly, if only in their dreams.

Discussion

The above dream FAQ addresses common questions about dreaming. As with all FAQs, this one is not exhaustive. The reader likely has additional questions. If so, some may be answered in Part Two, in which I apply emotional selection to deconstruct our most common dream themes.

People's fascination with dreaming dates back millennia. Aristotle and other ancient philosophers posited remarkably thoughtful explanations of dreaming. More recently, Freud's ideas dominated much of the discussion on dreaming in the early and mid-twentieth century. Following the discovery of REM sleep, researchers proposed several alternatives to Freud's. In the next chapter, I will review historic and contemporary dream theories and discuss how they fare in comparison to emotional selection.

SIX

The Competition

Entities should not be multiplied beyond necessity.

— William of Ockham, Ockham's Razor

MANY SCIENTIFIC FIELDS have competing theories. The field of dream research is no exception.

Theories of dreaming span a spectrum from religious to scientific, mystical to philosophical. However, their chronology is not as one might expect. Ancient civilizations produced surprisingly scientific theories, while many contemporary theories are unsupported by modern science.

Historical Theories

Plato's (427–347 BC) writings on dreams describe concepts familiar to modern psychoanalysts. He believed we harbor secret, repressed desires, such as incest, murder, adultery, and conquest, which build up during the day and run rampant during the night in dreams[55]—a concept that Sigmund Freud (1856–1939) embraced more than two thousand years later.

Aristotle (384–322 BC), a pupil of Plato's, developed a wholly different theory. He described sleep as sensory deprivation. Recognizing that our eyes are closed during sleep yet we still visualize during dreams, he concluded that dreams result from the mind interpreting nonsensical information from the eyes, ears, and other sensory organs.[56]

The father of Greek medicine, Hippocrates (460–377 BC), believed that dreams combined with astrological symbolism to reflect the physical health of the dreamer. A dream of a dim star high in the heavens symbolized difficulties in the head region, while low in the heavens represented bowel problems. He also believed that some dreams had divine origins, an opinion Aristotle contested.

Religions often describe dreams as having supernatural origins. The Bible interprets a Pharaoh's dream of the seven lean cows swallowing seven fat ones as signifying seven years of good harvests followed by seven years of famine,[57] and much of the Koran is said to have revealed itself to Mohammed in his dreams.

Contemporary Theories

The nineteenth century ushered in the new science of naturalism. Laypeople became enthralled with the study of nature, including the inner workings of the human mind. During this era, Freud formulated his ideas on dreaming, postulating that we repress immoral, socially unacceptable desires during the day that our mind must purge during sleep before they reach irrepressible levels. Because these violent scenarios are too overwhelming for our conscious minds to witness, Freud proposed that the mind disguises them with symbols, resulting in near-indecipherable dreams. So, the dreamer of a disturbing sexual fantasy may replace a penis with a phallic symbol, such as a tower or staff.

Freud theorized that interpreting dream symbols unlocked their hidden meaning, which he demonstrated with his interpretation of a client's dream:

> *She sees three lions in a desert. One of them is laughing, but she is not afraid of them. She must have run away from them, though, for she makes to climb a tree, but she discovers her cousin, a woman who is a teacher of French, already up there, etc.*[58]

Freud's interpretation:

> Her father had a beard that framed his face like a *mane.* Her teacher of English is called *Miss Lions.* An acquaintance sent her *Loewe's* ballads [*Loewe=lion*]. These are her three lions, then; why should she be afraid of them?—She has read a story in which a [slave] who has roused the others to revolt is hunted with hounds, and clambers up a tree to save himself. There follow, in a most playful spirit, remembered fragments, such as, "Instructions on how to catch lions" from *Die Fliegende Blatter* [A German humor magazine published in Munich from 1844 to 1944]: "Take one desert and pass it through a sieve, the lions will be left in the sieve." Then the very amusing, but rather improper, story of the official who is asked why he doesn't try harder to get into his boss's favor, and who replied that he has already tried crawling into it, but his immediate superior was already *up there.* The material all becomes intelligible when we learn that on the day of the dream the lady had been paid a formal visit by her husband's superior. He was very polite to her, kissed her hand, and *she was not at all afraid of him*, although he was such a big shot [in German idiom, *a big animal*], and was much *lionized* in her country. So, this lion can be compared to the lion in *A Midsummer Night's Dream*, who reveals himself to be

> Snug, the Joiner; and all the lions we are not afraid of in our dreams are just like that.[59]

Alfred Adler, a student of Freud, rejected his teacher's method of interpreting dreams. He believed that dreams were metaphorical rather than symbolic:

> Falling dreams … reflect concerns about falling socially and suffering a loss of prestige.[60] Dreams of flying are typical of forward-looking and ambitious people and leave the dreamer feeling buoyant and courageous because they portray overcoming difficulties as easy. Dreams of nudity or being scantily clad are associated with fears of exposure or being recognized as having an imperfection. Dreams of being paralyzed contain a warning that the present problem is without a solution. Traveling dreams refer to movement through life. Dreaming of a dead person indicates that the dreamer has not psychologically buried this person and is still influenced by them. If a person plays the role of spectator consistently in his or her dreams, this is probably the same role that they play in their waking life.

Carl Jung, another student of Freud, also diverged from his mentor. Jung believed the maturing mind unites the unconscious and conscious on its path to wisdom, and that dreams facilitate communication between the unconscious and conscious. Jung encouraged people to pay attention to the content of their dreams as a way of gaining insight into their inner growth. He believed that those who understood and embraced their inner struggles would find contentment.

Dream theorists of the late 19th and early 20th centuries, such as Freud and Jung, downplayed the physiology of the brain, believing that brain activity during sleep merely supported dreaming and other mental functions. In 1886, German physician Wilhelm Robert presented a notable departure. He proposed that our minds collect imagery and ideas during the day that must be "erased" to make space for new memories. Labeling this process as "dream-work," he considered dreams biological—an idea largely overlooked until the discovery in the 1950s that high-frequency brain waves often accompanied dreams.[61]

Several theories explain dreams as rehearsals for waking life. Much in the way athletes improve their skills through practice, these "practice play" theories describe dreaming as rehearsing for similar situations while awake.[62] In 1990, Jonathan Winson published his practice-play theory, which characterized dreams as providing an opportunity to relive past traumatic events to help prepare for similar future events.[63] In 2003, Antti Revonsuo introduced a similar theory, which suggested that dreams rehearse threats. So, a dream of being chased enables the dreamer to practice for threatening situations while awake.[64] In 2015, Revonsuo presented his second practice-play theory, which depicted dreams as a chance to practice social interactions, rather than threats.[65]

Our dreams often incorporate people and events from the previous day. This observation has led some researchers to describe dreams as a byproduct of consolidating new memories. Allan Hobson's *activation-synthesis* theory characterizes dreams as imagery resulting from the forebrain's attempt to make sense of the neural activity associated with memory consolidation.[66] According to Hobson, the forebrain attempts to understand these seemingly random activations, resulting in bizarre dream themes.

Researchers refer to theories that describe dreams as byproducts of brain processes as "brain-based." Conversely, "mind-based" theories, such as practice-play theories and emotional selection, hold that dreams are independent of memory consolidation and "housecleaning" operations performed by the brain. Rather than being byproducts, mind-based theories describe dreams as purposely constructed to serve a function.

Survival of the Sleepiest

A challenge for dream theorists is to explain why natural selection produced the processes described by the dream theory they favor. Borrowing this approach, we can explore the plausibility of the dream theories discussed above by considering which would have given dreamers an evolutionary advantage.

In *The Origin of Species*, Darwin summarizes the struggle to survive:

> We behold the face of nature bright with gladness, we often see superabundance of food; we do not see, or we forget, that the birds which are idly singing round us mostly live on insects or seeds, and are thus constantly destroying life; or we forget how largely these songsters, or their eggs, or their nestlings, are destroyed by birds and beasts of prey; we do not always bear in mind, that though food may be now superabundant, it is not so at all seasons of each recurring year.

Darwin reminds us that, beneath the beauty of nature, a fierce competition for survival exists. To improve the competitiveness of a species, its members often form groups. While there is safety in numbers, there is also intragroup competition. For instance, seagulls use calls to communicate. When a seagull sees a dog, it trills. Nearby seagulls repeat this call, alerting the entire flock to danger.[67] Observing this cooperation, one may conclude that seagulls always fend for the well-being of their fellow gulls. However, it only takes throwing a morsel of food into their midst to witness their fierce competition over limited resources.

Humans share this dichotomy of working for the greater good while simultaneously pursuing self-interests. We form search parties to find a missing child, donate our time and money to help homeless people, and are captivated by news of miners trapped by a tunnel collapse. However, when not preoccupied with the well-being of others, we can often be found competing for limited resources, including well-paying jobs, nice homes, and attractive partners.

People who strike a poor balance between competitiveness and cooperation often fare poorly in groups. Bullies are ostracized due to their inability to work with others, while naturally cooperative people who lack competitive instincts may be exploited. Balancing cooperation with competition increases a person's ability to succeed. However, how does one best strike this delicate balance? According to emotional selection, part of the answer lies in honing our social skills while asleep by adapting our schemas.

The social intelligence that emotional selection evolves during sleep is so essential to evolutionary success that the increase in human brain size is attributed, in part, to the increased capabilities of social intelligence. Simon Baron, a professor of developmental psychopathology at the University of Cambridge, summarizes:

> That there was a massive neuro-cognitive evolution during the Pleistocene epoch is beyond any doubt. The brain has increased threefold in size in the three million years since *Australopithecus afarensis* evolved, going from around 400 cubic centimeters to its current size of about 1,350 cubic centimeters.
>
> This increase in brain size is likely to have had many causes, but one key factor upon which many theorists agree is the need for a greater social intelligence short-

> hand that will give us the ability to process information about the behavior of others and to react adaptively to their behavior.[68]

The human mind is a remarkable byproduct of evolution, capable of solving complex intellectual problems. When considering our evolutionary success, we often think of tool making and our ability to adapt to varied environments as hallmarks. However, the evolution of our ability to form and navigate social groups is also vital.

The Expensive Brain

Cost often plays a role when making decisions. You want to buy that antique roadster, but that late-model sedan is more fuel efficient and reliable. You would love a tropical trip, but a "staycation" with friends and family is a better fit for this year's budget.

We use terms such as "expense" and "cost" when discussing money, but these labels also apply to the expenditure of limited or precious resources. For instance, naturalists characterize the evolution of ornate plumage on birds as "expensive" due to the energy required to grow large, colorful feathers.

Our brain accounts for two percent of our body mass yet consumes twenty percent of our energy, which makes the brain a very expensive organ. Consequently, prior to the discovery of REM sleep, researchers assumed the brain "hibernates" during sleep by shutting down nonessential mental processes to conserve energy. REM sleep revealed that, to the contrary, the brain is highly active during sleep. Instead of shutting down, its energy expenditure often surpasses levels during wakefulness, making brain activity during sleep an expensive process.

Natural selection disfavors species that waste precious resources. Since food energizes the brain, our prehistoric ancestors had to hunt and gather for this mental fuel. During our evolution, hunting and gathering could be dangerous endeavors. Natural selection selected high mental activity over mental hibernation during sleep. Therefore, dreaming and related mental activities are apparently worth the risk of acquiring the food to fuel an active brain during sleep.

The observation that dreaming consumes energy raises a challenge for dream theories that depict dreaming as a byproduct of unrelated processes during sleep. If dreams truly were useless byproducts, natural selection would have given the evolutionary advantage to members of species that evolved the ability to shut off the energy-hungry regions of the brain that process dreams.

Freud depicts a mind in conflict, in which some regions attempt to express themselves while others expend time and energy on repression. While plausible, Freud portrays a costly system that squanders precious resources on self-repression. Natural selection is averse to inefficiencies such as this.

Practice-play theories provide an explanation for why the brain is highly active during sleep, as rehearsing through dreams would have given our ancestors some evolutionary advantage. Meanwhile, emotional selection details a process that radically restructures and tests schemas to enhance their performance the following day. This restructuring of the workings of our mind during sleep would certainly be more expensive than practice-play and other processes described by mind-based theories, but it would also facilitate the meeting of essential waking needs. Naturalists and systems theorists like to see returns such as this on investments.

While all these theories are well considered, some describe a wasteful brain and mind. When contemplating the most probable dream theory, ask yourself, which proposal would give the dreamer an evolutionary advantage? Emotional selection describes a powerful process that radically reworks our schemas during sleep. Species that employed the less effective processes described by other theories would have fared poorly against species that modified and tested their schemas with dreams.

Discussion

This chapter concludes Part One. My aim thus far has been to explain how emotional selection enhances our ability to meet our needs. Armed with this knowledge, we can now comprehensively deconstruct dream content. Part Two begins with a simple approach to observing and logging your dreams. We then review the gamut of dreams your mind employs to meet your needs while you sleep.

Understanding the needs addressed by our dreams lends powerful insights into your waking life. You may feel stressed about work and money, yet dream of your personal relationships. You may believe that you overcame a past trauma, yet relive the trauma in your dreams. Reviewing your dreams helps identify the needs your mind is most concerned with meeting.

As we will discuss in the remainder of this book, deconstructing your dreams can provide insights that assist emotional selection in further improving your schemas. Let us now review the steps you can take while awake to understand and influence your dreams, and to better meet your needs.

Part Two

Your Dreams

SEVEN

Dream Deconstruction

Thoughts become perception, perception becomes reality. Alter your thoughts, alter your reality.

— William James

THE NEEDS that occupy our waking thoughts are not always the needs our mind works toward meeting while we sleep and dream. A single woman in her twenties, who participated in my online dating study, wanted a relationship, yet dreamed of financial insecurity:

After finding out that the owner of my home had sold it, I went to work to find out that one of my respected managers had lost her job due to money cuts and, whilst comforting her, was told I may be in the same position. The manager who I wanted to see would not see me as she was busy, but I could see through her window that she was just eating her lunch and reading magazines.

Although the possibility of losing your home and income is frightening, such concerns can distract our mind from working towards meeting higher needs, such as finding a life partner. I only collected a single dream per participant, so it is possible this woman often dreamed of attraction, just not on this night. Regrettably, despite wanting a relationship, she reported as still single in follow-up surveys.

Fortunately for this participant, and for anyone interested in using the insights of dreams, you can take steps while awake to shape your dreams, thereby shaping your future.

Dreams of finances and work are important, as they help us meet our need for security and cope when this need goes unmet. However, people who want to find life partners would be well served to also dream about relationships. But how does someone who excessively dreams of work improve their chances of dreaming about relationships? Simple—dream less about work.

Dreams of work often stem from fears of incompetence. These feelings of inadequacy plague even smart and successful people. If we dream about work because of a perceived sense of incompetence, we can evaluate our thought processes while awake by asking ourselves, "Do I make more mistakes than my peers?" "Do I produce less output?" "Am I less intelligent?" Often, such questions lead to realizations that our perceptions do not reflect reality. As our confidence in our work performance improves, we dream less about work and more about other aspects of our life, including relationships.

Facing Your Schemas

Unmet needs are disappointing. Being assigned a windowless office leads to unmet aesthetic needs, being unable to afford nutritious food leads to unmet physiological needs, and being accosted by a stranger results in unmet safety needs. While schemas typically cope with isolated instances of unmet needs, chronically unmet needs can damage schemas. For instance, a single slight from a loved one is upsetting, but repeated emotional abuse lowers self-esteem and perceived self-worth. Likewise, missing a paycheck may be challenging, but losing a job or enduring substantial financial losses damages related schemas.

Further complicating our ability to maintain healthy schemas is our tendency to ignore damaged ones. For instance, people who feel physically vulnerable may compensate by projecting an outwardly aggressive persona, or they may focus their energies on their career to wield power of a different sort. On the surface, such people may appear to be relatively free of fear. They may even associate their persona with strength, unaware that they overcompensate for perceived vulnerability. While they sleep, their safety dreams, such as being physically attacked or losing property, reveal their self-perceptions. Deconstructing such dreams unmasks valuable insights into schemas and brings maladaptive perceptions to the fore, allowing dreamers to face them directly rather than resorting to compensating strategies.

After identifying the needs targeted by your dreams, you can take steps while awake to improve your ability to meet those needs. For instance, recognizing that your dreams of safety recur, you can review whether your safety is truly threatened while awake; recognizing recurring dreams of esteem, you can review your perceived self-worth; and recognizing recurring dreams of belongingness, you can review your perceptions of your closest relationships. As you review the needs that appear in your dreams, you gain insights into the schemas that struggle to meet these needs while awake. Emotional selection can build on this newfound understanding to further improve your schemas while you sleep. Once satisfied that your schemas meet particular needs, emotional selection is then free to focus on other needs. In other words, deconstructing your dreams can be a step toward meeting the needs that appear in your dreams, resulting in a happier, more contented you.

Dream Opposites

Emotions have opposites, such as happy/sad, love/hate, and disgust/desire. Darwin wrote extensively about the pairing of emotional opposites:

> The tendency to perform opposite movements under opposite sensations or emotions would, if we may judge by analogy, become hereditary through long practice; and there cannot be a doubt that several expressive movements due to the principle of antithesis are inherited.[69]

Dreams also have opposites. Researchers Rosalind Cartwright and Lynne Lamberg observed that a single dream often contains opposite states or qualities, which they dubbed "dream dimensions."[70] For instance, dreams may contain themes of independence and dependence, trust and mistrust, and helplessness and competence.[71]

Dream theorist Patricia Garfield kept voluminous records of her dreams and those of others from around the world for more than fifty years. She assigned dreams to common themes, such as falling, flying, or being attacked. Like Cartwright and Lamberg before her, she recognized that dreams have opposites, describing dream categories as having unpleasant/pleasant themes. So, a frightening dream of falling is the opposite of an enjoyable dream of flying, a dream of being attacked is the opposite of a dream of being embraced, a dream of poor test performance contrasts with one of excellent performance, and so forth.[72]

According to emotional selection, pleasant dreams help us cope with met needs, while unpleasant dreams help us cope with unmet needs. We may dream of satisfying our need for safety by fending off an attacker. Other times, we fail to meet our safety needs as our nocturnal attackers prevail. While dreams of being triumphant are more comforting than dreams of being defeated, both types of dreams are essential for meeting our needs.

Deconstructing Dreams

Emotional selection integrates schema changes when they improve our emotional responses to a dream test. Upon waking, dreamers can assess the effectiveness of their schemas by reviewing their emotions during their dream.

Consider Julia's dream of being attacked:

> *I am standing in front of a closet. I see a black bug on a pretty, multicolored sweater. It is chewing holes in the fabric. I brush it off the sweater and it falls to the floor. At that point there is suddenly another woman standing next to me. She says to me, "You know, that's not a bug, it's a snake—and if you keep watching, you'll see it turn into a snake. Sure enough, as I stand there looking at the black bug, it starts to grow and turns into a big snake with shiny scales. As the bug grows into the snake, the woman standing next to me starts to shrink down and turns into a pretty little girl. I know that the snake is out to get us, so I grab the little girl, and we dash into the bathroom and shut the door to get away, but as soon as we get inside, I realize that even the bathroom isn't safe, because the snake can crawl under the door. No sooner do I have this thought than the snake does crawl under the door and into the bathroom. I jump up onto a little stool to get away from it, but there isn't any place for the little girl to get away, and the snake slithers up to her and rears up and bites her. I see this happening, and I jump down and start to beat the snake off the little girl, so that it falls to the floor.*[73]

In her dream, the snake was a threat. Julia found safety on the stool, but the child in her dream was still in danger and attacked by the snake. Instead of remaining on the stool, Julia risked her safety to protect the child. Her need to help the child took precedence over her need for safety.

At the time of her dream, Julia was coping with a trauma from her past; she was molested as a child. Fortunately, her schemas appear to be healing well, as she demonstrated in her dream that children are worthy of protection, even at the risk of an adult's life. Such perceptions are remarkable for a person who was unprotected as a child. The courage she displayed in her dream would certainly help her face her memories of her past abuse while awake.

Mark Twain said, "Courage is resistance to fear, mastery of fear—not absence of fear." Our schemas characterize our ability to muster courage. Self-perceptions of strength and worthiness help us fend off those who threaten us. Perceiving ourselves as ineffectual or unworthy of protection leads to self-perceptions that diminish our ability to act in times of crisis.

It is common for people to dream of scenarios vastly different from their waking life. Dreamers may live far from the ocean yet dream of being engulfed in a tsunami. Such dreams rarely reflect an overwhelming fear of a rogue wave. Rather, their minds are preparing to handle the chaos and crisis that accompany all disasters. Emergencies require fortitude and level-headed thinking. Emotional selection often enhances related schemas during sleep. Afterward, we dream of catastrophe to test these changes. While dreams of disaster may appear tangential to our waking life, the schemas tested are preparing for virtually any disaster we may encounter while awake.

Our broad spectrum of needs continues in our dreams. Imagine enjoying the company of friends when suddenly you sense what feels like a smooth pebble on your tongue. You discreetly remove it from your mouth and discover that it is not a pebble—a tooth fell out.

The next thing you know, your mouth feels full of pebbles, as all your teeth have likewise dislodged. What would your reaction be? Would you give your friends a wide, toothless grin and say, "Look, I don't have to bother brushing my teeth anymore!" Or would you be fearful for your health or mortified by your appearance?

The common dreams of suddenly discovering that your teeth fell out, your hair fell out, or that you suddenly aged often test schemas related to your physical appearance. Those with well-adapted schemas may feel initial dismay during these dreams, but then realize they possess other worthy attributes. They may say, "Well, my hair falling out doesn't make me any less of a good person," or "I may now suddenly be old, but I can still help others." Those who put too much stock in their physical appearance cope poorly with these dreams. Reviewing the thoughts and emotions we experience during such dreams lends insights into whether we place too much of our self-worth on too few of our attributes.

Several schemas are commonly shared among people and across cultures. Most of us possess the schema that survival is paramount, which explains why the common dream of falling from a great height is universally terrifying. However, schema networks vary between individuals, so our responses to dreams can likewise vary. For instance, the common dream of teeth falling out may trigger embarrassment for one dreamer and fear of mortality for another. Such differences highlight the importance of reviewing the emotions and feelings experienced during the dream when investigating the schemas tested. While dreaming of your teeth falling out is frightening, what was the nature of the fear? Was it public humiliation, aging, death, or something wholly different? Such variations in dream deconstructions reflect our individuality.

When deconstructing a dream, ask yourself, "Which situations during my waking life would elicit the same emotions I experienced during my dream?" A dream of forgetting to wear your clothes to work may evoke feelings of humiliation, but people rarely arrive to work unintentionally naked, so the dream does not prepare you for accidental nudity. Instead, the dream tested your ability to cope with humiliation in general. A more realistic waking-life equivalent might be humiliating yourself at your company's annual holiday party by doing something utterly embarrassing. If you drank too much at the annual event, grabbed the mic, and belted out a rendition of your favorite pop song, would you forgive yourself the next morning, or would the humiliation consume you? If a humiliating situation consumes you in a dream, you may be sensitive to humiliation while awake.

The most effective tool for deconstructing dreams is the simple dream log. Recording and categorizing our dreams reveals important insights into the feelings and emotions experienced during the dream, which helps identify the underlying schemas.

Logging Dreams

Our memory erases dreams roughly two minutes into non-REM sleep, so we typically remember only a fraction of what we dream. Fortunately, you can take simple steps to increase your chances of recalling dream details the following morning.

Immediately upon waking is the best time to record your dream, so keep a notepad or an audio recorder by your bedside. The longer you delay recording your dream, the more challenging it becomes to recall details, so jot down as many details as you can remember as soon as you wake. This also applies to when you wake during the night—do not postpone recording your dream until morning, as details will slip from your memory after falling back to sleep.

The feelings and emotions experienced during a dream are critical to its deconstruction. Do not allow the events of your dream to sway you into forming opinions you did not hold during the dream. A dream of being attacked may shift from terrifying to empowering if you turn the tide against your attacker. Likewise, a sex dream can be stressful if guilt or self-doubt plagues you during the dream. Dreaming is relative—a pleasant dream can be unfulfilling, while an unpleasant dream theme can have rewarding components. Understanding such subtleties lends insight into the performance of the schemas tested.[74]

Your bedside dream notepad or audio recording need not be eloquent. The goal is to capture the details of your dream for transcription into your dream diary the following morning. Based on my experience, this transcription should be done soon after rousing from bed, preferably within the first few hours, as transcribing the bedside recording often elicits dream features that you may have omitted in your bedside notes. So, transcribe your dream at the start of your day when the dream is still relatively fresh in your memory.

The template below helps deconstruct your dreams. You can use this template to transcribe your bedside notes into an organized dream log that captures the scenario of your dreams, the emotions experienced, and additional information essential for the deconstruction of your dream.

- **Date and Time:** Dreams occur shortly after falling asleep, shortly before waking, and anywhere in between. Record the approximate time of the dream.

- **Title:** Give your dream a title for reference in the future. Titles can be brief, such as "Ran into my ex."

- **Dream:** Record the events of your dream here. Do not elaborate on the feelings and emotions experienced during the dream—these are logged in the next step.

- **Emotions, thoughts, & feelings:** Record the feelings and emotions you experienced during the dream, such as anxiety, anger, joy, fright, happiness, and sadness. As mentioned, they may vary from the feelings and emotions you would have experienced if the events happened in your waking life, so do not be swayed by the dream's scenario. For instance, if you did something unethical during a dream but felt no remorse, note this.
- **Evaluation:** Describe how well your schemas performed during your dream. Hindsight is 20/20, so use it here. Did you behave bravely and admirably, or did you succumb to pressure? If you recognized the dream as a repeated theme, did your schemas perform better during this dream compared to previous ones?
- **Theme:** Dream scenarios may follow recurring themes that span months or years. Identify any such theme here. Use your own words, such as "This is another one of my 'being trapped' dreams." If this dream varied from your normal theme, such as being more extreme or surreal, make a note of this.
- **Pleasant or unpleasant:** Did your dream test your ability to cope with met needs (pleasant) or unmet needs (unpleasant)? This category can be tricky to identify. A pleasant dream of flying could induce anxiety if you struggle to lift off the ground, but it should still be categorized as pleasant.
- **Waking life influences:** Describe events during the day(s) leading up to the dream that may have influenced its content.
- **Need(s):** Which of your needs were incorporated into your dream? You can use Maslow's hierarchy or describe the needs in your own words. If uncertain, leave this blank or list possible needs, followed by a question mark.

- **Deconstruction:** Here is where you try to identify the perceptions and schemas your mind was trying to improve to better meet your needs. This entry need not be an essay—a sentence or two should suffice. For instance, "I need to trust my instincts more when protecting myself and others from being treated unfairly." If you don't know, leave this blank, as the dream may be the beginning of a new theme you have yet to identify.
- **Comments:** Miscellaneous observations or predictions go here, such as whether this is a new dream theme for you.

Dream deconstruction establishes how effectively the schemas performed that emotional selection targeted. A dream of being chased by zombies would be universally terrifying. However, the reaction to the dream would vary among dreamers. Some dreamers may act bravely by keeping their wits about them and helping others escape. Others may be overwhelmed, feeling unable to help themselves or others.

Below is an example of a dream report from my dream journal. It is one of my longer dream entries. To provide a backdrop, I enjoyed playing baseball and softball as a child and young adult and was a fan of my hometown baseball team, the Boston Red Sox. I have no association with the Red Sox aside from watching the occasional game on television.

- **Date & Time:** 1-Oct-2005, morning
- **Title:** Playing centerfield for the Sox
- **Dream:** I was at Fenway Park, working for the Red Sox during a playoff game against their rivals, the New York Yankees. My job was selling hotdogs or something to that effect. In the last inning, with the Red Sox leading 6–3, the manager substituted

four of us into the game to show his appreciation for our hard work, even though we were not professional athletes. I was horrified by the prospect of playing at such a pinnacle moment, but I did not question his judgment. Going out to centerfield, I couldn't find my good glove, so I used an old, broken one. After reaching centerfield, I found a wide niche in the home run wall that was a logical place to stand, only there was a problem—a picnic table on the field was in my way, forcing me to stand next to it. The Yankees seized the moment by intentionally hitting to the rookie players with unsurprising results—we made a lot of errors. They scored two runs quickly, making it 6–5. Then, Manny Ramirez, who plays for the Sox in real life but played for the Yankees in my dream, hit a line drive directly over the picnic table, right at the location where I had wanted to stand. I leaned over the table to make the catch, but between the old glove and the obstacle, I missed the ball. Picking it up as Manny rounded second base, I threw the ball from deep centerfield, but it was wide and he scored, tying up the game. The pitcher struck out the remaining batters, leaving the score tied and sending the game into extra innings. As I walked off the field, it occurred to me that not only was the game nationally televised, but that the whole Red Sox season likely would be summarized by my error. Had I caught the ball, we would have won the game, but since I dropped it, we might lose, making me the scapegoat for the 2005 season.

- **Emotions, thoughts, & Feelings:** I was okay with dropping the ball. Walking in, I was sad but also realized the manager was crazy for putting us in the last inning of such an important game. Because I did my best under the circumstances, and because even professional athletes drop difficult balls, it was not surprising that I did. Though, in hindsight, I should have refused the invitation to play.

- **Evaluation:** My need for acceptance appears to be in good shape. Even though I let down the team and the fans, I felt as though I did my best under the circumstances and shouldn't be overly criticized for that. However, my need for strength did not fare nearly as well. I should have trusted my instincts and declined the invitation to take the field. Rather, I folded to the will of the coach. I also used a beat-up glove when I should have held up the game until I found my good one. Likewise, I should have recognized the table as an unfair obstacle and insisted it be removed. I need to be more assertive.
- **Theme:** This follows the same performance/test dream theme that I've had most of my life, in which I must assert myself in the face of unfair obstacles. Usually, the dream occurs in a smaller setting. This is my first dream in front of a large crowd.
- **Pleasant or unpleasant:** Unpleasant.
- **Waking life influences:** The Boston Red Sox are about to start a playoff series against the New York Yankees. Also, my softball team just won the championship the day before, during which I played well, so I was feeling pretty good about my athletic skills. Maybe it was a good time to put related schemas to the test?
- **Need(s):** Strength, acceptance.
- **Deconstruction:** I still have issues with authority. I must recognize that people in positions of power often make poor choices, even when they have the best intentions. I need to be more vocal in rejecting direction when people are clearly wrong.

- **Comments:** This is the first dream like this in front of a public crowd. Usually, these dreams place me in front of small groups. Maybe my schemas are improving, and my mind turned things up a notch. I suspect this theme will continue in front of crowds until my mind decides my schemas perform adequately.

While playing baseball and softball during my younger days, I experienced a gamut of emotions, ranging from the joy of winning to the disappointment of losing. I also felt moments of intense pressure as the outcome of a game would occasionally fall on my shoulders or those of a teammate. Because of the important role these sports played in my life, baseball and softball games are occasionally the backdrop of my dreams. I have not played in years, but still find myself back on the field in my dreams.

Bizarre dream scenarios can make it challenging to determine the targeted schemas. By deconstructing the dream, recurring themes often emerge. I often find myself in a classroom during dreams, which test my ability to cope with humiliation and authority. I may take an exam and cannot get my writing utensil to work, or the room is too poorly lit to read the questions and I feel too intimidated to complain to the exam proctors. Until deconstruction, I did not realize that my Red Sox dream followed the same theme as my classroom dreams.

When recording your dream, be specific about the emotions, feelings, and perceptions of your dream. Dreaming of being chased by a three-headed monster does not mean you fear being attacked by a three-headed monster while awake. Such details in our dreams are surreal, in part because they are unimportant. Rather, the perceptions, emotions, and feelings of the dream relate its purpose.

Often, dreams contain themes with multiple needs. Consider the dream of Jasmine. Like Julia, whom we met earlier, Jasmine was coping with a sexual assault from her past. Rather than dreaming of her trauma, she dreamed of related, higher needs:

I was visiting some people on a remote farm. I found that a little girl there was being sexually abused. I rescued her, deciding that I would take care of her myself. As we were leaving, I found a large, incredibly beautiful jewel in her hair. We both knew it was mine.[75]

Jasmine's dream begins with the theme of rescuing a small girl, which Maslow describes as our highest need: transcendence. Once the girl reaches safety, the dream enters a different theme when Jasmine finds the beautiful jewel, which falls under another high Maslow need: aesthetics. Jasmine's real-life trauma damaged her schemas related to her basic need for safety. Rather than dreaming of her past trauma, she dreams of higher needs, suggesting that her damaged schemas are recovering.

A woman in her fifties also dreamed of multiple needs:

Vacation setting, mountains in the distance. It was evening, enjoying a seafood dinner. Comfortable weather, the smell of flowers blossoming. Wine, good conversation.[24]

Rather than treating a meal as mere sustenance, her dream integrated food with the satisfaction of other needs, including belongingness and her need for aesthetic beauty.

Maslow observed that people often simultaneously experience multiple needs:

> Eating may be partially for the sake of filling the stomach, and partially for the sake of comfort and amelioration of other needs. One may make love not only for pure sexual release, but also to convince one's self of one's masculinity, or to make a conquest, to feel powerful, or to win more basic affection. As an illustration, I may point out that it would be possible (theoretically if not

> practically) to analyze a single act of an individual and see in it the expression of his physiological needs, his safety needs, his love needs, his esteem needs and self-actualization.[76]

Maslow's description of how lovemaking can be more than "sexual release" appears in the dream of a woman in her fifties:

> *I was on a beach with this guy I've been dating. It was night. We were lying on a blanket, drinking wine and talking about what we like in a relationship, listening to soft music, then we started dancing under the stars. One thing leads to another, then we were making love, and when it was over we talked about it, and we both said it was the best and most open lovemaking we'd ever had.*[24]

In our lives, we often satisfy multiple needs simultaneously. The same observation applies to our dreams. This makes dream deconstruction both interesting and challenging. By isolating and identifying the needs of your dreams, you gain insights into the waking needs your mind works toward satisfying while you sleep.

How to Help Emotional Selection

As mentioned, after identifying the schemas in your dream that emotional selection improves while you sleep, you can take steps while awake to improve the schemas of your dream. So, you and emotional selection become a tag team—you work on your schemas while awake, and emotional selection picks up where you left off during sleep.

Sounds simple, right? Well, changing schemas can be challenging. As discussed, schemas often entrench in our schema network, particularly those developed as children. But, if up for a challenge, the benefit of improving your schemas is worth the work. What techniques are available to improve your schemas? A popular approach is cognitive therapy, which was developed by psychiatrist Aaron Beck.

Beck was a follower of Freud's school of psychoanalysis when he happened upon his discovery of cognitive therapy. During a session of free association—an exercise in which the therapist encourages patients to voice whatever thoughts enter their minds with no mental filters—a patient's free associations resulted in him verbally abusing Beck. Unfazed by the insults, Beck asked the patient how he felt.

"I feel very guilty," he responded. Beck initially thought the patient felt guilty because he insulted his doctor. Instead, the self-aware patient presented Beck with an interesting insight. He was not feeling guilty because of the insults, but because an internal stream of thought told him he was performing poorly: "I said the wrong thing ... I shouldn't have said that ... I'm wrong to criticize him ... I'm bad ... He won't like me ... I'm bad ... I have no excuse for being so mean."

Beck described the experience as "the first clear-cut example of a train of thought running parallel to the reported thought content." While the patient related a stream of thought, a second internal stream continually evaluated his performance. Beck wanted to know how common this second stream of thought was among his patients. During subsequent sessions of free association, he asked each to verbalize their previously unreported streams of thought. He found that they all had unreported thoughts, and that some were more aware of them than others. Beck relates his experience with a particular patient:

> A woman who felt continuous unexplained anxiety in the therapy sessions was describing certain sensitive sexual conflicts. Despite mild embarrassment, she verbalized these conflicts freely and without censoring. It

> was not clear to me why she was experiencing anxiety in each session, so I decided to direct her attention to her thoughts about what she had been saying. Upon my inquiry, she realized that she had been ignoring this stream of ideation. She then reported the following sequence: "I am not expressing myself clearly ... He is bored with me ... He probably can't follow what I'm saying ... This probably sounds foolish to him ... He will probably try to get rid of me."[77]

Cognitive therapy teaches us to recognize maladaptive thoughts and question their validity. For instance, when we question our competence or self-worth, cognitive therapy suggests we examine the evidence behind such thoughts. If self-doubt plagues us at work, we can ask ourselves, does our boss treat us differently than other employees? Do our performance reviews support this characterization? Cognitive therapy teaches us to challenge negative thoughts before accepting them.

Psychologist Jeffrey Young, a follower of Beck, noticed that cognitive therapy was effective on most of his patients, but some did not respond. He then changed tactics and focused on maladaptive schemas, instead of maladaptive thoughts, and achieved further progress. From his book, *Schema Therapy: A Practitioner's Guide*:

> The therapist and patient begin by agreeing to regard the schema as open to question. Rather than an absolute truth, it is a hypothesis to be tested. They then subject the schema to logical and empirical analyses. They examine the evidence supporting and refuting the schema in the patient's life; they go through the evidence the patient has used to uphold the schema, and they find al-

> ternative interpretations of these same events; they conduct debates between the "schema side" and the "healthy side"; and they list the advantages and disadvantages of the patient's current coping styles. Based on this work, the patient and therapist generate healthy responses to the schema.[78]

Young introduces Shari, a 28-year-old nurse and mother of two. Shari struggled with a self-schema of being defective, viewing the world through that lens. Among the related perceptions Shari listed were, "I'm fooling my husband and children, and that's why they love me. They don't know the real me. When people give me good feedback, I don't believe them. I think that there's some other reason they're saying it."

Shari's maladaptive schema developed during her childhood. Her father abandoned her and her mother when she was four years old. Her mother was an alcoholic who often humiliated her in public, including drunkenly attending her school play and interrupting it in mid-performance. By applying the techniques of schema therapy with Young, Shari addressed her maladaptive schema of being defective. Using techniques such as compiling written lists that challenged her perceived defectiveness, she saw abundant evidence that challenged her perceptions.

By applying schema therapy to your dreams after waking, you can scrutinize the accuracy of your schemas targeted by emotional selection. You may question your abilities, but after reflecting and comparing your skills to others, you may find they compare favorably. Scrutinizing perceptions helps nudge related schemas toward a healthier model of you and your world. Emotional selection can build on this momentum by further enhancing your schemas while you sleep. As emotional selection succeeds, you should see progress in your related dream themes, as well as improvements in your abilities to meet related needs while awake. Eventually, your recurring dreams should stop as emotional selection turns its focus to other schemas.

Schema flaws can originate from many sources, including having anxiety, being raised to be overly self-critical or critical of others, having a personality that gives rise to unproductive thoughts, or experiencing emotional trauma. Regardless of the source, reviewing your perceptions is a step toward improving the underlying schemas.

New perspectives kindle a refreshed self-image, which in turn enables us to gain novel insights about ourselves, others, and our lives. By viewing life from a different perspective, our schemas adapt and accommodate a new outlook. As they do, maladaptive schemas fade while healthier ones emerge.

Dream Themes

Recurring dreams have similar themes that often span months or years. The dream scenarios of recurring dreams can be identical, such as the same attacker chasing you down the same dark alley. Recurring dream scenarios can also vary. For instance, a common recurring theme is struggling with technical problems. You may try to dial a telephone and find it has no numbers, be giving a presentation at work and find yourself unable to plug cables into your laptop, or try to make a train connection and find yourself lost in a bewildering train station. Regardless of the scenario, the dreams involve the same feelings, emotions, needs, and schemas.

As the schemas targeted by recurring dreams improve, the recurring dream theme becomes increasingly tangential as emotional selection incorporates other needs. Once emotional selection is satisfied that the damaged schemas have healed, it moves on to other schemas.

Recurring dreams that do not evolve to incorporate other needs indicate a failure of emotional selection to improve related schemas. By applying the lessons of cognitive therapy, schema therapy, or another therapy of your choice, you can weaken the influence of your maladaptive schemas. Emotional selection can build on your progress to further refine your schemas while you sleep.

Consider the common dream theme of appearing incompetent. In the dream, you may give a presentation at work or speak before an audience and struggle with a malfunctioning microphone or video equipment. Variations on this theme include being late for a meeting or exam. Regardless of the situation, you feel anxious and try to correct what goes horribly wrong.

A man in his forties dreamed:

> *I had to get ready for work and I couldn't find any clean jeans—I searched all over, there was even a raffle for new clothes that I didn't win. I had to be at a meeting this morning and (the meeting was real) but I was two hours late because I could not find clean pants or any pants for that matter.*[24]

Inexperience and mishaps do not equate to incompetence. Even the most competent people among us make mistakes. Unfortunately, our schemas can be unforgiving. As you identify the inaccuracy of negative thoughts and feelings, you challenge the schemas that produce them, bringing to the fore those that reinforce general competence and self-worth. By challenging the accuracy of our schemas, we improve their ability to accommodate a new understanding of ourselves and our environments, which emotional selection can exploit to further improve schemas during sleep.

As your schemas improve, you will notice a progression in both the dream theme and your response to the dream. First, you will feel less anxiety, sadness, or other negative emotions and feelings because your schemas are better equipped to cope. Second, the themes may grow in complexity as emotional selection incorporates other schemas. Third, the intensity of the dream theme may increase as emotional selection taxes your newly improved schemas.

My Red Sox dream followed a theme that has recurred since I was a teenager. The scenario often varied, but the theme was the same—I am in a public situation where an authority figure treats the group unfairly, often unintentionally. Two years before the Red Sox dream, I had a dream that followed the same theme:

> *I arrived in a classroom to take an exam. Instead of desks, there is one bed for every student; all neatly aligned in rows. We each climb onto a bed, pick up a pencil and begin answering the questions. Lying on our bellies and propping ourselves up on our elbows, it is difficult to write, as the notebooks press into the soft mattresses. Things were manageable, however, until the lights slowly began to dim, eventually reaching a point in which the questions were difficult to read. My anxiety and agitation rose as I could no longer tolerate the unfairness of the situation. I finally spoke up and declared that the conditions were unreasonable. My classmates agreed and seemed relieved that someone had finally called attention to the problem.*

When I was younger, I hesitated to speak up during such dreams. My reticence mirrored my waking life, in which my need for acceptance from others inhibited my ability to assert myself. As I matured,

I developed a healthier perspective; I grew to understand the need to speak up when mistreated. As my schemas improved, I grew calmer during my classroom dreams. While these dreams were never enjoyable, I now found that I often asserted myself, which turned my unpleasant dream into a rewarding one.

My recurring dreams of coping with neglectful authority figures progressed from the relatively small scene of a classroom to the stadium of my Red Sox dream. My schemas improved their ability to cope with authority figures in a classroom of people, but could they cope with a stadium of passionate Red Sox fans? Unfortunately, my schemas did not perform admirably, as my baseball dream proved to generate too much anxiety, and I folded to the pressure of authority. I still occasionally have this recurring dream theme, as emotional selection continues to improve my ability to speak truth to power.

Ideally, emotional selection improves our schemas that support high self-esteem and self-worth. Unfortunately, the primary goal of emotional selection is adaptation, not happiness, as a diminished sense of self can be adaptive. For example, some parents overprotect their children to shelter them from physical and emotional harm. Displays of independence by such children may be met with resistance and fear from parents. Because the parent-child relationship is paramount, the dreams of an overprotected child may include themes that reinforce the child's vulnerability. So, such accommodations reinforce the parent-child relationship, but often lower self-esteem and self-worth.

As an example of a maladaptive dream scenario, consider a dream in which the dreamer is a perpetrator of violence. Beth, a woman in her thirties, dreamed of murder:

> *I dreamt that I murdered this woman and buried her in a shallow grave. I was walking around the area where I buried her with my ex-boyfriend and his daughter. I was*

> *really nervous that the dead woman was going to pop out of the ground in front of them because I didn't bury her deep enough.*[24]

Sometimes, even good people commit bad deeds. We may overlook a person in need on the street, fail to defend a coworker treated unfairly by a boss, or unreasonably snap at a loved one. Our schemas must be equipped to handle the spectrum of our behavior, both good and bad. During Beth's dream, her main concern was not remorse. Instead, she focused on not getting caught by her ex, expressing her concern that the corpse might "pop out of the ground." Her dream appeared to test her ability to cope with her immoral behavior while awake.

Regardless of whether our schemas performed admirably during a dream, we can review our dreams to gain deeper insights into the needs our mind tries to meet. We can then take steps while awake to improve the schemas of our dreams.

Discussion

In this chapter, we discussed how deconstructing dreams identifies the schemas that emotional selection strives to enhance. We also reviewed how the tools of cognitive therapy and schema therapy help to weaken maladaptive thoughts and the schemas that generate them. In other words, emotional selection can be used as a tool in conjunction with therapies to identify and improve maladaptive schemas.

Categorizing dreams by scenario (flying, falling, being chased) typically accentuates their most bizarre aspect: the scenario. Dreaming of being chased by a monster would never elicit feelings of joy. Rather, we feel the rational emotion of fear. The scenario is bizarre, while our perceptions and emotions are logical. What is needed are dream categories that highlight the rational aspects of dreams, not the bizarre ones.

In the next chapter, we will discuss a categorization of dreams that highlights the goal of emotional selection—meeting your needs. We then put the process of dream deconstruction via emotional selection to the test by reviewing common dream themes and their associated needs and schemas. As we do, you will likely recognize your dreams in the dreams of others. By deconstructing these shared dream experiences, you will gain insights into the dreams that shape your schemas.

EIGHT

Common Themes

Learn from your dreams what you lack.

— W.H. Auden

THIS CHAPTER is not a "dream dictionary." Dream dictionaries are texts that assign symbolic meaning to objects and events in dreams as a shortcut for their interpretation. So, water may represent a womb, resulting in a dream of drowning as representing an unconscious desire to return to infancy. Gayle Delaney, the author of several dream books, highlights the problem with interpreting objects as symbols:

> Just think of the ways that water can appear in a dream. You might be in a beautiful bath, drowning in a violent sea, or saving a child from a polluted stream. You might be drinking from a romantic Hawaiian waterfall or sailing in the San Francisco Bay. The water could be pleasurable or frightening, romantic or sexy, and it could take forms such as tidal waves or rain or snow. You might water-ski or swim with dolphins, play with sea otters, or look down the throat of a humongous white shark. You could be washing clothes or dishes in fresh soapy water or maybe even cold greasy water.[79]

According to emotional selection, a dream of drowning does not metaphorically represent drowning in the sea of life, nor does it symbolically represent a desire to return to the womb. Instead, a drowning dream tests the ability of your schemas to keep your wits about you when you are literally drowning. Similarly, a dream of a snake is not a phallic dream of sex, nor is it a metaphorical dream of dealing with sneaky, snake-like people. A snake in your dream is merely a limbless reptile.

Each person's schema network is unique. So, everyone's perspective of the world and themselves is distinctive. A dream theme that tests certain schemas of one dreamer might test wholly different schemas of another. You must deconstruct your dream to gain insight into the relationship between your dream and your waking life. This takes work. A dream often includes a gamut of emotions, which can make associating a dream with a single need or schema unsatisfying. Rather than shoehorn your dream into a single category, you deconstruct your dream by exploring all the emotions, needs, and schemas it triggers.

As discussed in the previous chapter, dream themes pair with opposites. So, an unpleasant dream of being threatened opposes a dream of feeling powerful, a dream of being incompetent opposes a dream of competence, a dream of being surrounded by ugliness opposes a dream of serene beauty, and so forth. In this chapter, each dream theme is paired with its opposite.

The Fine Print

Abraham Maslow's hierarchy of needs provides a convenient framework for categorizing dreams. His primary need categories (physiological, safety, belongingness and love, esteem, cognitive, aesthetics, self-actualization, and transcendence) can be further divided into subcategories. For instance, physiological needs can be further divided into survival, sustenance, and sex; belongingness and love needs can be divided into affection, acceptance, companionship, empathy; and so forth.

The techniques for deconstructing dreams, outlined in the previous chapter, apply to virtually all dream themes. As you deconstruct your dreams, you gain insight into your unique, individual needs and how emotional selection works while you sleep to improve the ability of your schemas to meet these needs upon waking. My choice of using Maslow's need categories is a personal preference. Dream deconstruction can be applied to virtually any system of need categories, including those devised by you, the reader.

As discussed, emotional selection modifies schemas primarily during non-REM sleep and tests schemas primarily during REM sleep. Because dreams are more common during REM sleep, and because we typically rouse from REM sleep, the dreams we remember upon waking are typically REM dreams. Since REM dreams are our most memorable dreams, we will discuss the dream deconstruction process below in the context of REM dream tests. This is writing shorthand, as the same principles that apply to deconstructing REM dreams also apply to non-REM dreams.

Within the sections of this chapter, I deconstruct dreams for each of Maslow's eight need categories. I begin at the bottom of his hierarchy (physiological) and work my way up to the top (transcendence). As I do, you may find commonalities in the dreams of others, and recognize the needs that emotional selection is attempting to meet while you sleep.

Physiological Dreams

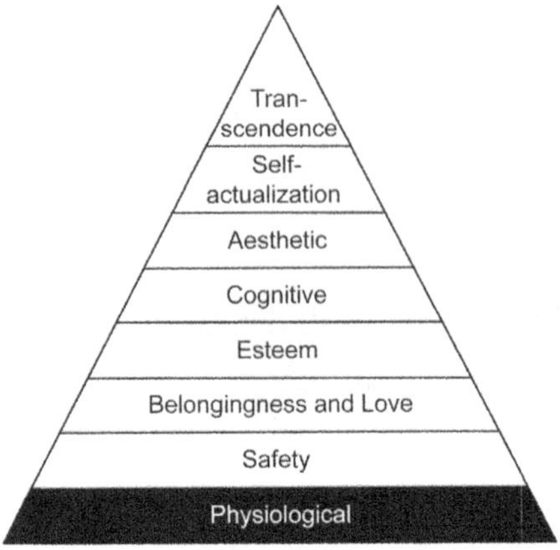

Physiological needs, which include our needs for food, water, air, sleep, and sex, occupy the base of Maslow's hierarchy. Unlike higher needs, such as our need for an aesthetic environment, physiological needs are often essential for maintaining life and cannot be deferred. We do not say, "I'm falling; I'll deal with this later." Instead, physiological needs typically must be met without delay.

Survival Themes:
Imminent Death Versus Supernatural Capabilities

When our life is threatened, we put every resource at our disposal toward meeting our need for survival. The ability of our schemas to cope with threats to our survival can mean the difference between life and death.

Like other dream categories, survival dreams can be both unpleasant and pleasant. During unpleasant survival dreams, we perceive that death is imminent. We may hide from a ruthless attacker or gasp for air as we drown. Regardless of the dream scenario, we believe that death is moments away unless we act. Conversely, pleasant survival dreams are often joyful. We may find ourselves free from the physical laws of nature, discovering we can fly, or that we possess supernatural powers of another sort. Both pleasant and unpleasant survival dreams are common, because our need to survive is paramount.

Thwarted Need for Survival: Dreaming of Imminent Death

In this terrifying dream, our life is in danger. We may be consumed by a tornado, struggle in a losing battle with a sadist, or fall from a great height. Variations include being trapped in muck or quicksand, or a tremendous weight pressing down on us, inhibiting our movement. We may attempt to cry for help and find our voice has stopped working. Regardless of the dream scenario, we believe we are about to die and are terrified.

Consider Harold's recurring dream:

> *I am fleeing in absolute terror, running through a barren landscape, pursued by something so horrible I cannot bring myself even to look back and see who or what it is. I run flat out through the sand and rocks with the horrible pursuer right behind me until I come to a big, deep chasm that blocks my path of escape. A frail-looking rope suspension bridge hangs across the chasm. I plunge onto the swaying rope bridge with the thought that, if I can just make it to the other side, I may be able to grab a quick look back and get a glimpse of who or what is pursuing me. But before I can get across to the other side, the rope bridge snaps, and I end up falling through space. This falling always wakes me up in terror, my heart pounding.*[80]

In Harold's dream, his schemas performed admirably. Despite his terror, he thought clearly and took the only logical course of action—run for his life.

A woman in her twenties had an equally terrifying dream:

> *I was swimming in a pool and had a swimming floatation belt on for water walking. It had a high back, and I put my head under in a weird way and could not get back up to the surface. I continued to sink and tried getting the attention of other people in the pool, [but it did] not work. I continued to sink and could not breathe, and I thought I would die. I woke up just as I felt myself slipping away. I woke up gasping for breath.*[24]

Dreams of unmet survival needs are universally terrifying. When schemas process these dreams effectively, we manage our fears and hide, flee, or resolve to go down with a fight. When processed ineffectively, we feel overwhelmed and helpless.

Your ability to respond to threats during dreams of unmet survival needs may improve over time as emotional selection adapts schemas toward meeting this important need. If you often dream of unmet survival needs without this progression, you can review your perceptions while awake. Perhaps you are recovering from past trauma, or a caregiver instilled the belief that you cannot fend for yourself. Regardless of your perceptions, identifying those that may limit your ability to manage your fears is a step toward enhancing underlying schemas. As your schemas improve, emotional selection can build on your progress while you sleep to help meet your need to survive.

Satisfied Need for Survival: Dreaming of Supernatural Capabilities

In this remarkable dream, we discover that we possess a unique power, such as superhuman strength or the ability to fly. We may take to the air, flying above cities and towns. Variations include other physical feats, such as unnatural swimming abilities or running at superhuman speeds. Regardless of the dream scenario, we perceive ourselves as above the physical laws of nature.

Heidi often dreams of flying:

> *I am outside, walking around, when I decide to leap and find that, to my amazement, I can glide in the air for a long time. I keep trying to leap farther and glide longer until finally I can actually fly up in the sky. I am shocked that I didn't realize earlier in my life that I had this ability. When I awaken, I am momentarily elated before I realize it was just a dream.*

Another woman in her forties dreamed of possessing superhuman powers:

> *Swimming underwater and was very warm and peaceful. I knew I was underwater, but I could breathe easily and deeply even though I was not wearing scuba gear.*[24]

Dreams of meeting our need for survival can be so invigorating that we want to return to our dream upon waking. Processing pleasant survival dreams without inhibition signifies our schemas are well-equipped to cope with times when our need for survival is met. When these dreams go awry, we can fly, but self-doubt impedes our newfound power as we drift back to earth, or our ability to wield another superpower is limited by a sense of unworthiness or a lack of confidence.

Recurring dreams of supernatural powers may grow more enjoyable over time as emotional selection adapts schemas toward embracing our essential need to survive. If you have recurring dreams in which you struggle to embrace supernatural powers, you can review relevant perceptions. Perhaps you feel unworthy of having your survival needs met, your parents were overprotective, or you endured a trauma that reinforced your mortality. Regardless of your perceptions, reviewing your ability to embrace pleasant survival dreams is a step toward enhancing related schemas. As these schemas improve during waking, emotional selection can continue your work while you sleep to further your ability to cope when your need for survival is met.

Sex Themes: Unsatisfying Sex Versus Satisfying Sex

Sex dreams often involve a gamut of schemas and perceptions. Maslow listed sex as a physiological need while acknowledging that it ties to other needs, such as esteem and love. Because sex dreams typically span multiple need categories, they must be carefully deconstructed to reveal the targeted schemas.

Like all dream categories, sex dreams have unpleasant and pleasant themes. Pleasant sex dreams test whether our schemas can fully enjoy sex, while their counterparts test whether our schemas can cope with occasions when our need for sex is unmet.

Thwarted Need for Sex: Dreaming of Unsatisfying Sex

Dreams of an unmet need for sex can take many forms. You want sex, but physical and social barriers, such as a jealous third party, make it difficult. You may find yourself in a public space without the requisite privacy, or continuous interruptions inhibit your ability to perform. You may engage in sex, but the act is unsatisfying as distractions or other problems bombard you.

Even sexually active people must cope with occasions when their need for sex is unmet. Dreams of thwarted sex test our ability to cope with the frustration and disappointment that can accompany the thwarting of this physiological need. (Dreams of sexual assault are discussed under Thwarted Need for Protection: Dreaming of Vulnerability, page 146. For dreams of being coerced into sex, see

Thwarted Need for Independence: Dreaming of Subjugation, page 189.)

Sandra said that her appetite for sex was much higher than that of her partner, Ron. This imbalance led to tensions and arguments. Eighteen months into their relationship, she dreamed:

> *At first, I was standing over my partner, Ron, wearing only a skimpy thong. We were on a bed that was soft and bouncy. I was laughing and jumping up and down … I felt fantastic and excited. I shouted, "Bounce with me!" over and over again. Then I noticed that Ron was looking odd. He seemed to be saying something but I couldn't hear him. I started to get angry, although I still wanted to make love to him. I lay on top of him with my legs straddling his … yet for some reason he couldn't respond. He was sinking deeper into the folds of the bed—I thought the covers were swallowing him up.*[81]

A woman in her sixties also encountered challenges while attempting to have sex:

> *[I] dreamed I had met this man in his camper and we were about to have sex. But his son and other people kept showing up at like 5:30 a.m. with a horse trailer and walked right into the camper. I was in the bed with no clothes on, and he got up to visit with his son. I watched as I lay in the bed in the camper as others drove up and he talked with them. Finally, a woman showed up that he appeared to have had or was having a relationship with. She saw me and smiled and started to talk with him as if I was not there. I popped up and asked her to leave so we could have sex. She laughed at me and stated she was his girlfriend and she was not leaving. The dream became broken as I walked around a house which*

> *he appeared to own with her with the camper just outside the back door. People just kept showing up and he would talk to them and ignore me. I finally found myself getting dressed and heading for the door. Several dogs that appeared in the yard seemed to like me and I bent down to pet them. At that moment, I woke up.*[24]

Dreams of an unmet need for sex are often discouraging. When our schemas process dreams of thwarted sex effectively, our sexual attempts fail, but we do not believe it reflects on us. We manage to keep our emotions in check as we embody the understanding that not every sexual urge can be met. When processed ineffectively, we become overwhelmed or overly frustrated. We may feel that we are somehow at fault or undeserving of meeting our sexual needs.

The emotions that accompany dreams of thwarted sex may improve over time as emotional selection adapts schemas toward meeting this important need. If you often dream of thwarted sex without a progression, you can review related perceptions while awake. Perhaps your personality has trouble coping with unmet needs in general, or barriers exist in your waking life that limit your ability to meet your need for sex. Regardless of your perceptions, assessing those associated with sex is a step toward resolving conflicts among underlying schemas. As these conflicts are resolved, emotional selection can build on your progress while you sleep to further improve your ability to cope with times when your need for sex is unmet.

Satisfied Need for Sex: Dreaming of Satisfying Sex

In joyful dreams of sex, we find ourselves in the arms of an attractive partner. He or she may be a past lover, current boyfriend or girlfriend, spouse, or complete stranger. Regardless, there are no interruptions and minimal inhibitions.

A woman had a pleasant sexual dream that included her platonic friend, Dave:

> *I dreamt that I was meeting Dave for a night out. I was rushing to get ready and get over to his house. When I arrived, he was wrapped in a towel. I was anxious and told him to hurry up. Once he was dressed, I led him into the bedroom, saying we were going to have a great night "out" (even though we were in). He was hesitant, and I was saying, "Come here or we'll be late." The next thing I knew, we were on his bed … [having] very vigorous sex. Dave's flatmate then walked in and said, "Oh, it's you two," and carried on talking to us as though we were on a regular evening out.*[82]

The dreamer had few inhibitions during her sex dream, suggesting her schemas processed her dream well. She even appeared unphased by the interruption, describing the flatmate as behaving as though nothing unusual was occurring between her and her friend Dave.

A woman in her fifties, who preferred not to engage in sex during a date, dreamed of bending her rules and meeting her sexual needs:

> *My dream was of a sexual encounter. I had invited a man over for dinner. Before we could even eat, things got very sexual.*[24]

The woman remarked that her dream was unusual because she typically prohibits her dates from "getting sexual." Based on the content of her dream, her schemas adapted to meet her physiological need for sex.

When schemas process sex dreams effectively, the sex is satisfying and free of inhibition. When related schemas perform poorly, our self-image or perceptions stifle our ability to enjoy sex. These feelings stem from our inhibitions, not from our partner, who remains committed to the act.

Pleasant dreams of sex may grow increasingly enjoyable over time as emotional selection enhances related schemas. If your dreams of sex could be more satisfying, you can review your perceptions while awake. You may find you have insecurities or attitudes toward sex that limit your ability to perform in both the dream and waking life. Regardless of your perceptions, reviewing your ability to enjoy sex is a step toward improving related schemas. As your schemas improve, emotional selection can build on your progress while you sleep to improve your ability to embrace your need for sex.

Food, Warmth, and Other Physiological Themes

Maslow recognized that physiological needs are our most primitive. Even in today's modern societies, people's most basic needs often go unsatisfied. Those who struggle to put food on the table may have unpleasant dreams of their need for sustenance being thwarted, such as a dream of futilely searching for food, or may dream of seeing their need for sustenance met as emotional selection conjures a banquet in a beautiful setting.

The satisfaction of our physiological needs is essential, which is why Maslow placed them at the base of his pyramid, and why emotional selection adapts the schemas responsible for meeting these important needs while we sleep.

Safety Dreams

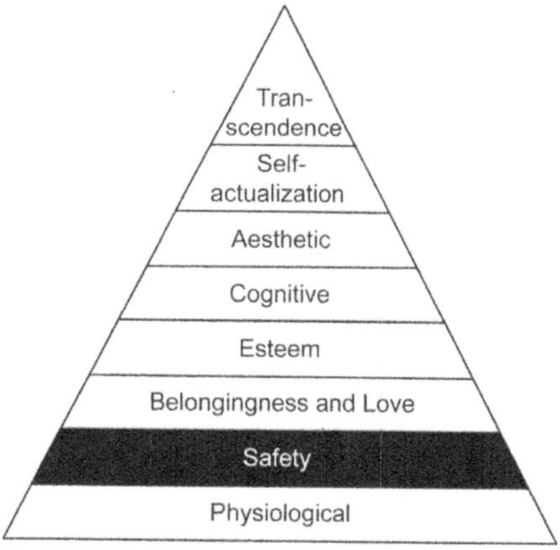

Maslow described our safety needs as our need for organized, calm, and secure environments. As a society, we create laws and employ governments to ensure our safety. As individuals, we introduce order and routine into our life. Although people embrace change, they also value safety—a need so essential that Maslow placed it immediately above our physiological needs in his hierarchy.

Refuge Themes: Disaster Versus Tranquility

Maslow characterized our need for refuge as the need for "protection from the elements." We all seek environments free from danger. For instance, people travel great distances to vacation in tranquil surroundings that evoke a sense of calm, and beautiful vistas of the ocean, fields, and other open spaces are peaceful, in part, because no harm is within sight.

Emotional selection ensures that our schemas are equipped to cope with our met and unmet needs for refuge. So, we have unpleasant dreams that place us in hazardous scenarios to test our ability to cope with times when our need for refuge is thwarted, and pleasant dreams that place us in tranquil settings to test our ability to cope with occasions when our need for refuge is satisfied.

Thwarted Need for Refuge: Dreaming of Disaster

In this stressful dream, a natural or human-made disaster thwarts our need for refuge. The natural disaster may be a flood, tsunami, earthquake, or hurricane. The human-made disaster may be war or a collapsing building. Regardless of the dream scenario, we are in danger and must act quickly.[83] (For dreams of losing valuable property, causing financial anxieties, see Thwarted Need for Financial Security: Dreaming of Lost Property, page 142. If faced with leading other people to safety, see Thwarted Need for Efficacy: Dreaming of Powerlessness, page 185. If death is imminent, see Thwarted Need for Survival: Dreaming of Imminent Death, page 129. If the dream is frustrating, rather than dangerous, see Thwarted Need for Stability: Dreaming of Chaos, page 150.)

In his waking life, fire almost destroyed Michael's house. Three weeks later, he had a disaster dream:

> *In my house watching the flood waters rise. Soon I look out the window and the ocean waters are coming right up to the edge of the house. I start getting worried. Ultimately, the waves start splashing against the house and water starts leaking in. I take my computer, unplug it, and start to carry it upstairs. More water starts to leak in. I tell my wife we may get flooded and she gets worried. Then as I watch out the window, the flood recedes like in the movie The Ten Commandments. Mary, a friend who has bad fire damage, is wet and upset. I quickly receive two tennis warm-up jackets from friends to help me recover.*[84]

In his dream, Michael took more measures to protect his computer than he took to protect his family. Later, he appears to look to others for refuge, receiving jackets from his friends and making no mention of whether his family reached safety. Although his schemas did not spur heroic action, they were effective in securing his survival.

Real-life trauma, such as Michael's experience of nearly losing his home to fire, can trigger recurring disaster dreams as emotional selection continually repairs related schemas. Feeling overwhelmed in disaster dreams signifies our schemas may be ill-equipped to handle an emergency. In times of crisis, we may be consumed by fear, making it difficult to improvise a plan of action. Saving ourselves but not feeling empowered or motivated to save loved ones could be a sign that we hold our personal safety too dearly and may be unprepared to protect others.

A woman in her thirties dreamed of disaster in the form of a tornado:

> *I was at work and went outside to smoke. Once out there, I saw a tornado coming. I ran inside and yelled to everyone. I grabbed a hold of a wood shack, but the pieces were being ripped apart from the tornado. My friend, Nate, was being pulled into the sky by the winds, and I grabbed him and pulled him down. The only thing that helped us from being pulled into the eye of the tornado was a paperweight.*[24]

Dreams of disaster can be terrifying. When schemas effectively process dreams in which our need for refuge is unmet, we focus on protecting ourselves and others by locating an escape route or sanctuary. When schemas perform poorly, our ability to think or act is clouded.

The fears experienced during dreams of thwarted refuge may lessen over time as emotional selection adapts our schemas. If you have recurring dreams of refuge and feel helpless or overwhelmed during these dreams, you can review related perceptions. Perhaps you were raised to look toward others for refuge, or your personality incorporates a heightened sense of mortality. Regardless of your perceptions, reviewing those that inhibit your ability to meet your need for refuge is a step toward enhancing related schemas. As your schemas improve, emotional selection can further build on your progress while you sleep to enhance your ability to cope with occasions when your need for refuge is unmet.

Satisfied Need for Refuge: Dreaming of Tranquility

In this tranquil dream, our need for a safe environment is met. We may stroll through a beautiful field, admire a view from a mountaintop, or explore a sparkling city. Regardless of the dream scenario, natural or human-made splendor surrounds us. Everything is wonderful, and we feel at peace.[85] (Dreams of discovering new rooms or spaces are described in Satisfied Need for Knowledge: Dreaming of Learning, page 198.)

A woman in her fifties dreamed of tranquility:

> *I was in a pleasant place, I don't know where, and I was happy and carefree and all my wants and needs were taken care of. Someone was doing any and everything I needed before I even knew what I wanted myself. I was in total peace and as happy as I can ever remember being.*[24]

Author Patricia Garfield relates an elderly woman's recurring dream:

> *Several white doves fluttered against a sunlit window, with shafts of light flooding the dining room. She was careful not*

to open the door to let the birds out. She awoke happy and full of loving feelings.[86]

Relaxing in peaceful surroundings recharges our batteries. Calming dreams indicate emotional selection is working toward enjoying refuge and peace in our waking life. When processed effectively, we perceive ourselves as safe in our dream, free from danger. When processed poorly, we may enjoy our surroundings only briefly, finding ourselves distracted by concerns about unseen perils. Although life can be dangerous, being overly focused on danger can be debilitating. Accepting that some events are beyond our control may help us embrace tranquility.

Dreams of meeting our need for refuge may grow increasingly pleasant over the course of recurring dreams as emotional selection reworks related schemas. If you often have these dreams and struggle to enjoy their tranquil moments, you can review your perceptions while awake. Perhaps an influential person in your past fixated on potential dangers, or your personality is overly concerned with hazards even in their absence. Regardless of your perceptions, identifying those that limit your ability to relish moments of tranquility is a step toward enhancing underlying schemas. As your schemas improve, emotional selection can build on your progress to further develop your ability to embrace tranquility.

Financial Security Themes: Loss of Property Versus Acquisition of Property

In contemporary society, financial security is often synonymous with safety. Money pays for shelter, clothing, and food. So, the loss of money or property can challenge our ability to meet other basic needs. Emotional selection may modify schemas related to financial security and test these changes with unpleasant dreams of losing something valuable or pleasant dreams of finding treasure.

Thwarted Need for Financial Security: Dreaming of Lost Property

In this unsettling dream of financial loss, a precious possession may be stolen, an expensive belonging vanishes, or a natural or human-made disaster destroys our home. Common variations include a missing wallet, jewelry, or another valuable possession.[87] (For dreams of lost items that trigger perceptions related to knowledge, such as forgetting where we parked our car or being lost, see Thwarted Need for Knowledge: Dreaming of Ignorance, page 197. If loss of property results in losing status, see Thwarted Need for Status: Dreaming of Embarrassment, page 174.)

Marjorie had the common dream of losing her money:

> *I was walking downtown, and I suddenly noticed that I had lost my purse. I don't know if it had been stolen by a pickpocket or if I had dropped it, but I realized I had no money left with me at all, and I started looking everywhere. I realized that I would never find it.*[88]

Belongings aid in our sense of safety, but can be typically replaced when lost or destroyed. Losing a wallet or other personal items is distressing, but typically not catastrophic.

A woman in her forties dreamed of a grave financial loss:

> *I had lost my job and I couldn't feed my son. We had nothing. I was selling things to buy food. Then I lost the house, and we (were) in the street watching men board our house up. That's where I woke up.*[24]

Dreams of missing or destroyed property range from uncomfortable to distressing. When our schemas process dreams of financial loss effectively, we maintain perspective. People with robust perceptions of security typically see losing a job or money as a temporary setback. If the loss in the dream is small, we realize it is not the end of the world. If the financial loss is significant, well-adapted schemas set to work formulating a plan on how to recover.

The stress experienced during recurring dreams of unmet financial needs may lessen over time as emotional selection adapts schemas toward coping with the thwarting of this important need. If your finances are healthy and you often dream of losing valuables, you can review related perceptions. Perhaps you were raised in a household in which possessions were overly valued, or you feel pressure to keep up with the Joneses. Regardless of your perceptions, deconstructing dreams of unmet financial needs is a step toward enhancing underlying schemas while awake. As your schemas improve, emotional selection can build on this progress while you sleep to further develop your ability to cope with challenges to your financial needs.

Satisfied Need for Financial Security: Dreaming of Acquired Property

In this enjoyable dream of acquiring property, we may find treasure or win a lottery. The valuable may be something in nature, such as a gold nugget or a precious gem, or we may discover that we own property that had escaped our notice, such as an expensive automobile or a rare antique.[89] (For dreams in which a new acquisition leads to an increase in status rather than financial security, see Satisfied Need for Status: Dreaming of Prestige, page 175.)

A woman in her sixties dreamed:

> *Dreamed I won the lottery and was able to help my children and the people I used to work with. I was able to provide for some folks that live down the street from me who are in*

> *bad shape, as both parents are unemployed. I built my dream house and woke up just as I was going to open the door to my dream house.*[24]

A woman in her fifties also dreamed of winning money, which led to unexpected challenges:

> *I went to Vegas with friends. I played a $.05 machine with my friend, Steve. He doesn't like to gamble, and I told him to bet 15 nickels. He thought that was too much. Made comments. Anyway, he doubled his money on the first spin. I remember looking up at a big screen and seeing 15 different slot games being played, and when you won there was a blue box reflecting the number of nickels you won. I got him going, and I went to the machine across the aisle. I played the max, and I remember seeing the second box with 28,652. I won that many nickels. I was really excited. A guy came by and said I had won $862,000. I was so excited. Steve was pissed, so he walked away. They were working on getting me my check. I told the security guy I would be down the aisle on another slot machine. I walked down past a bar and saw another machine like the one I won on, and I put in a $20 and bet the max 15 nickels. Didn't win. Bet again. Didn't win. On the third time, I won $5.2 million dollars. I was kind of embarrassed. I remember looking at the security guard by my other winning slot, and he just shook his head. I remember thinking that I won $7 million in less than one hour.*[24]

Dreams of acquiring property are often joyful. When our schemas process these dreams well, we do not see our newfound wealth as life-affirming. Rather, we experience genuine joy. When processed poorly, our initial excitement fades as feelings of unworthiness and other concerns come to the fore.

Dreams of acquiring property may grow increasingly enjoyable as emotional selection adapts schemas toward embracing literal good fortune. If you frequently dream of meeting your financial needs and find that you are unable to embrace dream moments, you can review related perceptions while awake. Perhaps you have difficulty satiating your need for wealth, or feel unworthy of financial security. Regardless of your perceptions, assessing your ability to embrace financial security during waking is a step toward enhancing related schemas. Emotional selection can build on this progress while you sleep to further improve your ability to enjoy times when your need for financial security is met.

Protection Themes: Vulnerability Versus Dominance

Our need to defend ourselves is both primitive and important, which explains why we often dream of our need for protection. These dreams may include primal themes that bear little resemblance to waking life. For instance, people living in urban settings may dream of being pursued by a lion, or those living in land-locked areas may dream of being hunted by a shark. The source of the danger in our dream is unimportant, as many of the same schemas responsible for protecting us are triggered regardless of the type of threat.

The pleasant version of protection dreams tests whether our schemas embrace the empowerment of feeling protected. As we shall see, these dreams often include a dominance theme that tests whether we struggle with dominating those who would do us harm.

Thwarted Need for Protection: Dreaming of Vulnerability

Dreams in which our need for protection is unmet are typically frightening. We may be chased, hunted, or attacked. Our assailant could be a person, animal, monster, or other threat, or we may sense the presence of an attacker rather than see him, her, or it. We may find ourselves wedged into a tight space or sinking in quicksand, crippling our ability to escape. Our pursuer may try to kill, rape, or gravely injure us. Dreamers often report they somehow know the motives of their attacker without being explicitly told.

Violent attacks can occur from strangers or even once-trusted family members. Victims of violence suffer damage to their schemas that emotional selection attempts to repair. Even those of us fortunate enough to avoid trauma in our waking lives may find ourselves subjected to violence in our dreams, as emotional selection enhances the ability of our schemas to cope with the possibility of violence in the future.

Heidi, who shared her dream of flying earlier, also dreamed of vulnerability as a child:

> *I am on the swing set in my backyard. I keep glancing nervously over my shoulder, sensing something was behind me. Finally, I see a bunch of gorillas climbing over the fence. I run inside the house. My heart pounds as I watch the gorillas reach the house and bang on the door.*

Heidi's childhood dream shows she is acutely aware of her need for protection. Rather than succumbing to her pursuers in her dream, her schemas performed admirably, enabling her to master her fears and escape danger.

A woman in her twenties, who had been attacked in real life, had a relevant recurring dream:

> *I was trying to get away from someone chasing me. They caught up to me and pinned me to the floor, immobilizing me. This is a recurring dream, and when I usually have it, I don't struggle and just give up. Last night, however, I did struggle and managed to break free. That's never happened in the dream before.*[24]

This woman's dream demonstrates the progression that appears in dreams. Her improved schemas were no longer overwhelmed, as she mustered the courage to struggle and break free.

The primary goal during dreams of attack is to manage our fears. When schemas process attack dreams effectively, we keep our wits about us and escape, hide, or confront our attacker. We may initially feel terror, but push through our fears as we fight for our life. When processed ineffectively, we feel overwhelmed and incapable of protecting ourselves.

The anxiety experienced during dreams of an unmet need for protection may lessen over time as emotional selection adapts schemas toward meeting this essential need. If you often dream of attacks and your effectiveness in coping with this dream theme is not progressing, you can explore perceptions you hold in your waking life that contribute to your feelings of vulnerability. You may recognize that you are recovering from a trauma and consequently feel overwhelmed when threatened, or that you rely on others for protection. Regardless of your perceptions, deconstructing your dreams of vulnerability is a step toward enhancing related schemas. While you sleep, emotional selection can build on your progress to further develop your ability to meet your need for protection.

Satisfied Need for Protection:
Dreaming of Dominance

In this empowering dream, we protect ourselves and dominate a dangerous situation. The source of the threat could be a natural disaster, a monster, or an assailant. Regardless of the scenario, as the dream theme unfolds, we or our protectors are stronger than the threat and prevail. (For dreams of supernatural capabilities, such as super strength, see Satisfied Need for Survival: Dreaming of Supernatural Capabilities, page 130.)

In protection dreams, our protector may be of this world or another. A woman in her fifties dreamed of being protected by angels:

> *I dreamed that my 35-year-old daughter and I were in a second-story apartment with windows all around. We looked out the window and saw a tornado far off. We began to watch and soon two more tornadoes appeared, one on each side of the original. A glance out of a window in a room on the other side of the apartment noted that there were as many tornadoes on the other side of us as well. The tornados continued to come closer and increase in size until we were completely surrounded with all of the tornados being side by side. My initial reaction in a time of crises is to pray Psalms 91 that God will place angels in charge over us to keep us in all of our ways. Angels appeared shoulder to shoulder between us and the storms. We could see the storm going on through the angels but we were completely protected and calm. The funniest thing is that I remember a tiny airplane buzzing around somewhere high up above us which made me think that we were in the eye of the storm. I felt completely serene knowing that everything was okay. I rarely remember dreams, but this one I remember well.*[24]

I rarely dream of dominance. However, I dreamed:

> *I witnessed a man dragging a woman into his car. I somehow knew he was going to rape her. She struggled and screamed, but was no match for him. I pulled him from the car, threw him to the ground, and beat his face with my fists. Even though he became unconscious, I decided to continue the beating, literally beating his head off. When I stopped, there was nothing where his head once was. I then lifted his body and hurled it onto his car, scribbling a note and pinning it to his clothing that said, "This is what happens to rapists."*

In my aggressive dream, I felt justified in killing the attacker. Rape is a heinous crime. However, it is not typically punishable by death. I became the attacker in my dream, and the rapist was now a victim. Two weeks later, I dreamed of another attacker. Toward the end of this latter dream, I wrestled my opponent into a chokehold, but showed restraint. I eased my grip, sparing the attacker's life. My schemas appear to have improved during the interim, allowing me to wield dominance responsibly.

Dreams of dominating an enemy feel empowering. When our schemas process these dreams well, we use our power effectively and with restraint, wielding it only as far as the situation requires. When processed poorly, we struggle to protect ourselves and others, or perhaps bully our enemies, showing little mercy.

The emotions experienced during recurring dreams of dominance may progress toward calmness and other positive emotions as emotional selection enhances schemas toward managing power. When dominance dreams lack progression, you can review related perceptions. Perhaps an influential person in your life would feel threatened by your empowerment, or possessing power conflicts with your sense of self. Regardless of your perceptions, identifying the underlying schemas that emotional selection targets is a step toward improving your ability to assume power. As your schemas improve while awake, emotional selection can build on this progress to further develop your ability to wield dominance responsibly.

Stability Themes: Chaos Versus Order

Life is unpredictable. We may be enjoying a morning coffee one moment and navigating screaming kids, an angry boss, or a computer virus the next. Because our schemas must cope with unpredictability, we dream of when things go smoothly and when things go awry. So, like other dream categories, stability dreams have an unpleasant theme (chaos) and a pleasant theme that is its polar opposite (order).

Thwarted Need for Stability: Dreaming of Chaos

In this distressing dream of thwarted stability, we attempt routine tasks, but things invariably go sideways. We may use a computer that continuously malfunctions, or drive a car with nonsensical controls. Regardless of the dream, our attempts to rectify our predicament fail, making it impossible to meet our goals. This dream embodies Murphy's Law—anything that can go wrong will go wrong.[90] (For dreams of ineptitude, see Thwarted Need for Adequacy: Dreaming of Ineptitude, page 181. For dreams of feeling powerless, see Thwarted Need for Efficacy: Dreaming of Powerlessness, page 185.)

A woman in her thirties shared a recurring dream:

> *I visit my childhood home quite often in dreams, since my mother lost it due to foreclosure back in 2005. The house is always just as I've always known it to be: furniture in its*

> *normal place, paint on the walls same as when the house was ours, etc. Not always the same people in the dreams, though. I'm always there, yes, but sometimes it's just me and my kids; or my mother, stepfather and me; or, just me and my niece, sometimes my brother, mother and me. The people always vary. However, whenever my mother [and] stepfather are there, there is always fighting and chaos. Many times, police cars [are] out front. Same sort of dreams, about three times a week. Very rarely do I dream about anything else. But these dreams are very livid, lucid. A lot of memories in that house.*[24]

Dreams of chaos often feature themes of traveling or commuting. Psychologist Fionna Starr relayed her patient's dream, which included the mundane experience of automobile traffic:

> *Joel dreamt he was driving a brand-new sports car around his local neighborhood on the way to collect his girlfriend. It was a hot summer day, and Joel was driving with the roof down. He felt really proud of his new possession and was enjoying the attention that he was getting for it from passersby. Suddenly, Joel hit a massive traffic jam on a small road. He waited patiently, but after five minutes of not moving, he started to get frustrated. He began tooting his horn to see if it would make any difference, but to no avail. Joel then decided in his dream that it would be easier to turn around and take another route to his girlfriend's place. He managed to turn in a drive, only to find that the traffic going the other way was just as bad, and he was stuck for another ten minutes that seemed like hours. By this time, Joel was incredibly frustrated, and he ended up smashing his car into*

> *the car in front. He reversed and did it again and again, until his rage had, in fact, destroyed his own vehicle that he was so fond of.*[91]

During dreams of chaos, things go wrong. We have a goal, but external forces work against us, interfering with our objectives, which causes seemingly endless anxiety and frustration. When our schemas struggle during chaos dreams, the emotions can be unbearable, as demonstrated by Joel's dream. When our schemas process dreams of thwarted stability well, we maintain perspective and remind ourselves that chaos is a part of life that we occasionally must endure.

While dreams of chaos are never pleasant, feeling overwhelmed during chaotic dreams may signify related schemas are ill-equipped to cope with chaos. If you often dream of an unmet need for stability, you may find that you are increasingly able to weather chaotic experiences in your dreams. When recurring dreams of chaos show no such progression, you can assist emotional selection by reviewing your perceptions while awake. You may realize you have an overdeveloped need for order, or you feel unable to mitigate chaos. Regardless of your perceptions, as you learn to cope with chaos while awake, your related schemas should improve. Emotional selection can build on this progress while you sleep to further enhance your ability to cope with chaos.

Satisfied Need for Stability: Dreaming of Order

The antithesis of dreams of chaos is dreams of order. During these dreams, we or those around us may effortlessly operate a car, boat, plane, or some other equipment. Alternatives include systems or procedures that execute flawlessly. Regardless of the dream scenario, we relax as everything proceeds smoothly. In a stability dream, life is good, and we feel calm.[92]

A woman in her forties dreamed of contentment while traveling:

> *[I am] riding on the back of a motorcycle. I do not know the driver, but the wind is in my hair and I am winding up the side of a mountain. I can smell the fresh air and the trees and I am at peace.*[24]

A writer also dreamed of travel:

> *I am on a big oceangoing tugboat covered over with glass. I am part of a group. We are heading into a storm, mountainous waves fifty feet high. It looks dangerous, but the captain knows exactly what he is doing: [he]slices right through the waves. The boat remains fully upright, never slows down, and takes the storm beautifully. I marvel at the sturdiness of the craft and the skill and confidence of the captain.*[93]

This dream of being on a ship in a violent storm could have been stressful, but was instead enjoyable. When our schemas process dreams of satisfied stability well, the surrounding orderliness pleases us. When processed poorly, we feel anxious, believing the smooth operation is transient and that chaos will ensue.

Recurring dreams of order may grow increasingly satisfying over time as emotional selection enhances related schemas. If you find that you are unable to embrace dream moments during stability dreams, you can review your perceptions while awake. You may realize you are pessimistic, believing chaos is looming behind every corner and that order is the exception rather than the norm, or you may find chaos overwhelming, inhibiting your ability to enjoy its absence. Regardless of your perceptions, deconstructing your dreams of order is a step toward enhancing related schemas. Emotional selection can build on this progress while you sleep to further improve your ability to enjoy occasions when your need for stability is met.

Personal Boundaries and Other Safety Themes

Our personalities dictate which experiences we perceive as threatening. A morally rigid person may feel threatened by immorality and consequently dream of others and themselves acting immorally. People who perceive themselves as physically or psychologically frail may dream of physical or psychological threats.

Safety needs are a broad topic that includes our need to protect ourselves and others. The same principles of emotional selection that apply to all need categories also apply to safety needs—an unpleasant dream, in which our safety is endangered, tests our ability to cope with thwarted safety needs, and its pleasant, polar opposite tests our ability to embrace occasions when our need for safety is satisfied.

Belongingness and Love Dreams

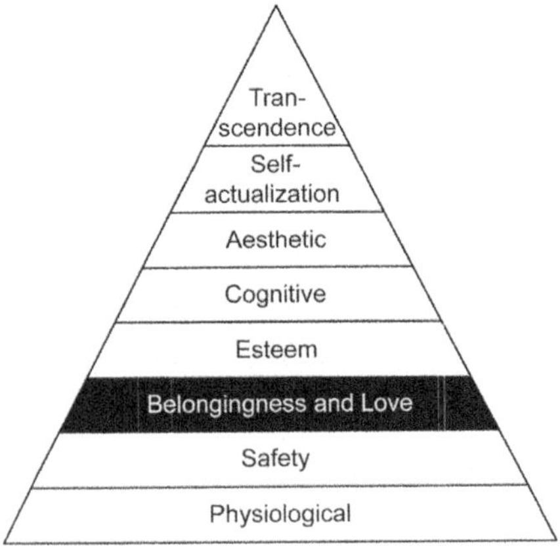

The physiological and safety needs at the base of Maslow's hierarchy ensure humans and animals survive to procreate. If the lower needs are sufficient for the perpetuation of species, why did we evolve to need belongingness and love?

As usual, the answer is evolutionary—our need to form and maintain meaningful relationships helps us meet more basic needs. For instance, there is safety in numbers. Our higher need to form social groups helped our ancient ancestors fend off attacks, indirectly meeting their need for safety.

In modern society, people often specialize in meeting the needs of others. Carpenters, contractors, architects, electricians, and other building specialists provide our shelter. Grocers and farmers supply our food. People intuitively form friendships and seek spouses because our personal relationships help us meet other needs. We do almost anything for those we love, as no need of theirs is too small. In turn, those in our life pay it forward by helping others meet their needs.

To meet belongingness and love needs, our schemas must prepare for entering and maintaining meaningful relationships. We must genuinely care for others and feel deserving of care in return. Feeling unworthy of love or of belonging to a group can foster mistrust, dread, or anxiety, which can sabotage our relationships. Emotional dreams, such as being accepted or rejected by loved ones or groups, modify and test our ability to meet our need for belongingness and love.

Affection Themes: Rejection Versus Adoration

Few things in life are more satisfying than giving and receiving genuine affection. Mutual expressions of caring enhance our self-esteem and strengthen relationship bonds. A nurtured child typically matures into a well-adjusted adult, equipped to handle life's successes and failures, while a neglected child can develop chronic levels of anxiety and insecurity, becoming quickly frustrated during difficult times and less likely to appreciate good times.

Our schemas must cope with the thwarting of our affection needs, which is why we have unpleasant dreams of being separated or rejected by loved ones. Our schemas must also cope with the satisfaction of these needs, which is why we have pleasant dreams of receiving adoration from those we love.

Thwarted Need for Affection: Dreaming of Rejection

In this unsettling dream, we lose an important source of affection. A person or group may reject us. We may blunder and turn to others for forgiveness or emotional support, and no one steps up. A close friend, family member, or lover may refuse our embrace. We may find that our spouse or partner abandoned us for another, leaving us for a friend or co-worker. Variations include someone close to us dying.

Pearl, whose boyfriend was uncommitted emotionally to their relationship, dreamed of rejection and separation:

> *Gilles and I were in my bedroom. He got out of bed, went to the door, and waved goodbye. I knew he was leaving me for a sexy blonde, and I cried myself awake.*[94]

When our need for affection is unfulfilled, we may feel empty inside. Pearl said she knew of the problems in her relationship and where they were likely to lead. While she slept, emotional selection prepared her for losing a significant source of affection in her life. Her dream was a dry run of the rejection she expected, which tested whether she was emotionally equipped for the anticipated breakup.

A woman in her twenties dreamed of her real-life coworker, whom she secretly loved.

> *I dreamt that we were hanging out together, and we were holding hands and doing what couples do. Then his wife*

> *showed up, and he left me. And I was devastated. It was a dream that has that overwhelming negative connotation to it, so I woke up and was super sad.*[24]

In dreams of rejection, we lose a source of affection, which saddens us profoundly. When our schemas process this dream well, we feel sad, but believe the loss does not reflect our character or our ability to enter future loving relationships. We view the scenario as a byproduct of unfortunate events, rather than as reflecting our self-worth. When our schemas process this dream poorly, we perceive ourselves as flawed and undeserving of affection, and consequently deserving of the loss.

Heartbreaking dreams of rejection should grow more bearable over time as emotional selection adapts related schemas. If you often dream of rejection without this progression, you can review your perceptions while awake. Perhaps your parents or caregivers had difficulty expressing love during your childhood, or you endured the emotional trauma of a breakup. Regardless of your perceptions, identifying those that make you overly vulnerable to rejection is a step toward healing underlying schemas. As your schemas improve, emotional selection can build on your progress while you sleep to further develop your ability to cope with rejection.

Satisfied Need for Affection: Dreaming of Adoration

In this heartwarming dream, we feel emotionally close to someone. The person may be a family member, a significant other, a friend, a baby, or even a pet. We may embrace, hold hands, or simply chat. We may dream of being reunited with a deceased loved one, such as a parent, sibling, or spouse. Sometimes the dreams are sexual, but the sex is secondary to the closeness we feel. In the dream, we sense acceptance and love. Upon waking, we may become sad when we realize the experience was only a dream.

A single woman in her twenties dreamed of the adoration of a future family, complete with an adoring husband and new baby:

> *I dreamed that I was pregnant; that I could feel the baby in my stomach, even feel it kick. The dream took me through the pregnancy, classes, shopping for baby stuff, doctor appointments, everything. There weren't any specific events or incidents in the dream and not a lot of interaction with other characters in the dream. It was mostly just me going around doing what I normally do, only I'm pregnant. In the dream I am working in my desired occupation, I am in a happy relationship with a guy I have not yet met in real life, and I am very happy and content.*[24]

Marla was a sixty-five-year-old woman who dreamed of her husband three years after he died:

> *I dreamt I was sitting in the living room when, suddenly, Jonathan walked in. He looked so healthy and happy. He put his arms around me, and we embraced. It was so wonderful to be with him; it was so real, it felt much more real than a dream. He said that he couldn't stay very long; he just wanted me to know that he loved me. Then the dream was over.*[95]

In losing her husband, Marla lost an important source of affection. In her dream, emotional selection tested her schemas with a powerful source of love in the form of her deceased husband. Her schemas processed her dream well, suggesting she can cope with loving again.

Accepting love from important people in our life is essential for meaningful relationships. When our schemas process adoration dreams well, we share a mutual exchange of affection with someone special. Unfortunately, people often perceive themselves as unworthy of giving or receiving adoration. They may shy away from others, act prickly to keep people at bay, or display an impersonal veneer to the world. Such people may desire closeness, but are ill-equipped to give or receive it.

Emotionally warm dreams adapt our schemas toward incorporating more affection into our life. If you experience anxiety during these dreams, you can review related perceptions that may cause you to withdraw emotionally. Perhaps you feel undeserving of affection, or are suspicious of overt demonstrations of closeness, believing others attempt to control you. Regardless of your perceptions, gaining insight into your ability to give and receive affection is a step toward improving the underlying schemas. As you work on your schemas while awake, emotional selection can build on your progress while you sleep to help you embrace affection in your life.

Acceptance Themes: Ostracism Versus Integration

Social groups play a significant role in fulfilling our needs. When working with others toward a common goal, we simultaneously meet our needs and theirs. However, groups have their downsides. Even caring, socially skilled people can be subjected to bullying and ostracism from others.

The opposite of being rejected by a group is being accepted. So, we often dream of ostracism and its polar opposite, integration, as emotional selection adapts our schemas to cope with the social side of being human.

Thwarted Need for Acceptance:
Dreaming of Ostracism

In painful dreams in which our need for acceptance is unmet, we find ourselves isolated. We may be the new kid moving into a neighborhood, back in high school trying to join a sports team, or on our first day at a new job. Regardless of the dream scenario, we try to fit into a new group and fail, or we may already be a member of a group, such as a family, workplace, or team, and others distance themselves from us. Variations include being separated from our group physically because of an impassable obstruction, such as a ravine or great distance, or our friends moving away, leaving us alone. In yet another variation, we dream of dying and attending our own funeral, in which we realize we may never see our loved ones again.

A woman in her twenties dreamed:

> *I had accidentally killed someone I went to school with, and everyone wanted me to go to jail. But, the police kept saying it was an accident, so I was fine but everyone hated me.*[24]

Julie was a thirty-two-year-old woman with marital problems and an eating disorder. Her weight loss resulted in diminished productivity at work. During this period, she dreamed:

> *I was dreaming about one of my former co-workers, that she was getting ganged up on by her fellow employees. Apparently, she was sick and tired of the job, and was wanting to quit. And then it turned into a video of her. Instead of her being there in person, we were all sitting around, watching it. She was crying and perspiring and laughing and obviously kind of losing it, mentally, in front of all of us. I was enraged. I was turning around and yelling at some of the people behind me. Rather than saying, "Look, I think she*

> *needs some help," it was like they were getting a kick out of watching this movie. This was supposed to be at work, but it was in my parents' living room.*[96]

Dream researcher Rosalind Cartwright interviewed Julie about her dream:

> Julie acknowledged that she was the one who felt ganged up on. Her productivity fell at work when she became ill and was trying to get out of her bad marriage. She assailed her former coworkers, whom she felt did not offer her support. She felt they merely pretended to be helpful, but actually stood back and watched her self-destruct and even got a kick out of her dilemma.[97]

Upsetting dreams of ostracism adapt our schemas for integrating into groups. When our schemas process these dreams well, we consider ourselves an outcast, but do not believe the circumstances reflect our character. Instead, we feel we contribute to groups, believing that others benefit from having accepted us. When processed poorly, we feel undeserving of being a group member. In the dream, we may give up and flee, preferring solitude over rejection. Rather than attributing the experience to outside influences, we feel that we alone are the problem.

Distressing dreams in which our need for acceptance is unmet help schemas cope with ostracism. If you often have these dreams, you can review related perceptions while awake. Perhaps you have anxiety, or your caregivers instilled a belief that you lack the skills to meaningfully contribute to a group. You may feel continually judged by others, even without signs to support this idea. As you assess the contributions you make to the groups in your life, you take a step toward enhancing related schemas. Emotional selection can build on this progress while you sleep to further improve your ability to cope with times when you feel ostracized.

Satisfied Need for Acceptance: Dreaming of Integration

In this enjoyable dream, we feel accepted by others. The source of acceptance is a group, such as our friends, coworkers, or family. We may enjoy laughs together or work side-by-side toward a common goal. Regardless of the dream scenario, we find ourselves openly welcomed and appreciated. (For dreams of being loved, see Satisfied Need for Affection: Dreaming of Adoration, page 157.)

A man in his twenties shared that he did not make time for important things in life. He dreamed of integrating with his family and community, including those with whom he had difficult relationships:

> *[The dream] was very relaxing because it was just a nice hike around a lake close by. I went swimming then went to work, but saw so many people that I know aren't even close by. It was nice, even the people I didn't get along with were my best friends, and the family members who detest me welcomed me.*[24]

A woman in her fifties dreamed of integrating with a local, real-life indigenous group whom she admired and longed to be with:

> *[I] saw myself going up the mountain, met the group, explored what could I do to help them—maybe education, maybe a cooperative, maybe market their products, not sure how, but they welcomed me. [They were] like my family, making my life more meaningful and happy.*[24]

Gratifying dreams of meeting our need for acceptance are often heartwarming and joyful. When our schemas process integration dreams well, we bask in belongingness. When processed poorly, we may feel elated initially, but subsequently experience self-doubt or dread as we question our worthiness. The group accepts us, but we absorb little of their intentions as self-consciousness consumes us.

Recurring acceptance dreams may grow increasingly rewarding as emotional selection improves the ability of our schemas to embrace acceptance from others. If you often dream of acceptance without this progression, you can reconsider your schemas and the perceptions they produce. Perhaps you have roadblocks in your waking life that inhibit your social success, or that you believe that you are flawed or continually judged by others, even when no evidence supports this conclusion. You may realize that an important person in your life, such as a parent or spouse, feels threatened by or jealous of your popularity. Regardless of your perceptions, reflecting on your ability to integrate with others is a step toward enhancing related schemas. As your schemas develop during waking, emotional selection can work while you sleep to further adapt your ability to embrace acceptance.

Companionship Themes: Isolation Versus Closeness

Loneliness is mentally and physically unhealthy. Psychologists from the University of Chicago found that people who were lonely perceived themselves as more helpless and threatened. In addition, lonely study participants had less restful sleep, resulting in fatigue and reduced cognitive function.[98]

Solitude does not equate to loneliness. Introverts are comfortable with occasional moments of being alone, with only their thoughts to keep them company, and philosophers, artists, and spiritual readers have described the virtues of spending time with oneself.[99] However, even the most devout believers in solitude may occasionally long for the company of others, which contributes to why emotional selection adapts our ability to cope with isolation, and with its polar opposite, closeness.

Thwarted Need for Companionship: Dreaming of Isolation

In this sad dream of isolation, we suffer the loss of friends or family. The setting could be a room full of people, yet we feel alone, unable to connect in a meaningful way with others, or we may find ourselves alone in a literal sense, perhaps in an empty house or in the wilderness. Regardless of the dream scenario, the absence of loving people in our life leaves us with a profound sense of loss and loneliness. (Dreams of being lost in the literal sense are described in Thwarted Need for Adequacy: Dreaming of Ineptitude, page 181. Dreams of being rejected are discussed in Thwarted Need for Affection: Dreaming of Rejection, page 156.)

A woman in her twenties dreamed of the isolation that accompanies the loss of a parent:

> *I dreamed my father had died and I was completely devastated, as it happened so quickly, I never got to say goodbye. The feeling of emptiness was so real, I knew I was crying in my dream. However, toward the end of my dream, it had changed to my mum who had died. I dreamed of feeling very lonely and depressed and scared.*[24]

I dreamed of the loneliness that accompanies the loss of friends:

> *I was back at university and my last roommate just moved out. I realized that I had no roommates left and felt very lonely. I was taken aback by how I had let this happen. Steve*

and Paul were my roommates, but they were gone now. My room accommodated four beds, but I was the only one there. There were leaves on the floor and I tried to sweep them up, but all I had was a mop, which wasn't the right tool for the job, so wasn't working well. Then, the floor turned to grass. I wasn't quite sure what to do next. I put my mind toward coming up with a plan to get roommates, but also knew that whatever I came up with would take a lot of time.

The people in my dream were my friends at university. While the dream felt real, it did not reenact a past event. However, during this dream, I experienced the most profound loneliness I had ever felt, awake or asleep.

To share closeness with another person requires emotional vulnerability. People occasionally isolate themselves from others to avoid the pain of rejection. Although isolation can spare us from loss, separating ourselves from others also limits the potential for meaningful relationships. Closeness is both precious and rare. To feel close to someone, we must also prepare to cope with loss.

Disheartening dreams of profound isolation test our ability to cope with the downside of making ourselves vulnerable to others. When our schemas perform well during these dreams, we understand our situation is unfortunate, but also feel it does not reflect our character. When our schemas perform poorly during isolation dreams, we feel unworthy of closeness, and believe that isolation is our destiny.

The discomfort that accompanies dreams of an unmet need for companionship may lessen over time as emotional selection works toward meeting this important need. If you dream of isolation without this progression, you can review your waking perceptions for those that may inhibit relationships with others. Perhaps your parents or caregivers shunned closeness, or you recently lost a loved one and may want to avoid another such loss. Regardless of your perceptions, assessing those related to closeness while awake is a step toward enhancing underlying schemas. As your schemas improve, emotional selection can further enhance your ability to weather loneliness.

Satisfied Need for Companionship: Dreaming of Closeness

In this joyful dream of meeting our need for companionship, we experience a personal connection with others. The people may be family, friends, acquaintances, or strangers conjured in our dream. The exchange may be platonic or romantic. We may bond with an individual or several people. Regardless of the dream scenario, as we exchange intimate emotions and feelings, we share a kindred spirit. (Dreams in which a group accepts us with little or no emotional exchanges are described in Satisfied Need for Acceptance: Dreaming of Integration, page 162.)

Despite not planning to attend his twentieth high school reunion, a man in his thirties dreamed of the closeness he shared with his classmates:

> *I had a dream that I was at my 20th high school reunion (even though I vowed in real life not to go). It was great seeing everyone. The reunion was a positive experience in the dream.*[24]

In the 1940s, a young college woman dreamed of a reunion of a different kind—her brother's return from war. She shared that her brother had called recently and spoke with her parents when she was not home. She was excited he called, but said that she was glad she missed his call because, for reasons we do not know, she would not have known what to say. She later dreamed:

> *In my dream someone tells me that my brother is on his way home and is planning to drop in on us unexpectedly. My mother, sister, and I decide to surprise him and be at the station to meet him. When we arrive, we see my brother and his boyfriend, also enlisted in the U.S. Army. They seem to see us, but pass right by. My brother then turns around, takes out a flashbulb camera, and takes our picture. I then remember my mother embracing my brother and crying. I run up to my brother, hug him, and with tears welling up into my eyes, keep repeating that I am so glad he has finally come home. I remember that as I was hugging him, a little voice within me kept saying, "See, you didn't have to worry about how you'd act when you saw Steve."*[100]

Her dream tested her ability to cope with expressing feelings toward her brother. Based on her outpouring of emotion, her schemas performed well, as she had no inhibitions.

Uplifting dreams in which our need for companionship is met are often accompanied by heartwarming love and contentment. When our schemas process these dreams well, we may initially feel reluctant to express our feelings, but eventually embrace our emotions as we reunite with loved ones or simply hang with friends. When processed poorly, we may long for closer relationships in our dream, but struggle to express our emotions.

The stirring emotions that accompany dreams of closeness may grow over time as emotional selection adapts related schemas. If you struggle to fully embrace dream moments when your need for companionship is met, you can review your perceptions while awake for those that may limit your relationships. Perhaps you fear that others will not reciprocate your feelings or that emotional exchanges interfere with other aspects of your relationship, such as a sense of authority held by you or someone else. Regardless, reviewing related perceptions is a step toward connecting with others. As your schemas adapt during waking, emotional selection can build on your progress to further improve your ability to maintain close relationships in your life.

Compassion Themes: Indifference Versus Empathy

A study of people caring for terminally ill dependents demonstrates the power of compassion. The dependents had doctors whose demeanors ranged from indifferent to empathetic. The researchers found caregivers working with indifferent doctors were more likely to be depressed and were more likely to report that taking care of their dependents interfered with their personal lives.[101]

Compassionate people are often at ease entering new relationships, inherently knowing they have a powerful, personal gift they can bestow upon others. The importance of embracing moments of compassion in our life is why we dream of empathizing with others. However, even compassionate people may occasionally act with indifference. So, we must also cope with occasions when we lack empathy, which is why we may also dream of acting indifferently toward others.

Thwarted Need for Compassion: Dreaming of Indifference

In this unsettling dream, we lack compassion. People in our dream may have lost a loved one, been ostracized, or simply made a trivial mistake. We understand their situation, but are indifferent toward their feelings. (Dreams in which we are compassionate but thwarted from helping people are described in Thwarted Need for Altruism: Dreaming of Failing Others, page 213.)

Chuck was a young Ph.D. student in the physical sciences who dreamed:

> *A group of friends, which included Chris and Casey (a married couple), and I are standing around in my living room. There is a pause in the conversation, and Chris says, "Everybody! Casey has an announcement to make." All eyes turn to Casey. After hesitating for a second, Chris blurts out, "Casey and I are going to have a baby!" I am surprised by the announcement and exclaim, "Oh, I am so happy for you!" I turn to Casey, who happens to be standing next to me, and give her a big hug. "This is great news," I say to her and Chris but, actually, I have very mixed feelings. I am concerned that I will be spending less time with my friends and drifting apart from them.*[102]

The gamut of emotions Chuck experiences in his dream mirrors the complexity of the human condition. His friends' good news left Chuck with mixed feelings. He wants to be happy for their good fortune, yet feels conflicted and sad because of his potential loss of their companionship. Had Chuck felt only jealousy toward his friends, his feelings would suggest that his schemas fostered indifference. Instead, his complex feelings showed that his schemas support his relationships.

Chuck was a relative bystander to the events of his dream unfolding before him. However, our dreams may also depict us behaving with marked indifference. Betty, a woman in her fifties who was facing financial stress in her waking life, dreamed of being uncharacteristically mean:

> *I was at a pool; the pool looked stagnant. I passed by Sue (a coworker and friend) and kicked her hard in the back of her calves and made her fall in. I walked away, perplexed why I had done it; she is so sweet and had not done anything to deserve my anger. I turned around and apologized for*

> *"accidentally kicking her" and others helped me pull her out. She was upset. No one else had seen what had happened, but I knew and was torn mentally. My other coworkers thought, poor little Sue and rallied around her.*[24]

Dreams of indifference can be unnerving. When our schemas process these dreams well, we view our thoughts or actions as callous, but also understand that others are worthy of compassion. When processed poorly, we feel entitled to act inappropriately, placing the blame for our actions on others.

The callousness that accompanies recurring dreams of indifference hopefully diminishes over time as emotional selection adapts related schemas. If you occasionally dream of indifference and your dreams are not progressing, you can review related perceptions while awake. You may realize that your caregivers did not value compassion when you were a child, or that they ignored your emotional needs. Alternatively, you may find you place undue emphasis on self-reliance, perceiving compassion as unnecessary, which limits your ability to connect with others. Assessing whether your schemas harbor indifference is a step toward improving the underlying schemas. As your schemas improve, emotional selection can build on this progress while you sleep to further enhance your ability to navigate times when you feel indifferent.

Satisfied Need for Compassion: Dreaming of Empathy

In this emotional dream, we empathize with others. They may have won a competition, received a beautiful gift, or been promoted. Regardless of the dream scenario, we share in their joy. Alternatively, people in our dream are enduring hardship, and we comfort them by sharing in their grief. During dreams of empathy, we experience a deep connectedness with others as we experience their emotions as though they were our own. (Dreams in which we bond with others are described in Satisfied Need for Acceptance: Dreaming of Integration, page 162. Dreams in which we are compassionate and help people are described in Satisfied Need for Altruism: Dreaming of Helping Others, page 215.)

Barb dreamed of empathizing with her brother, Jake:

> *Jake is driving. We are in the M City back roads area and probably going to his house. He's all locked in and driving fast and I say, "Hey, come on. If you want to commit suicide, go ahead, but don't take us with you." He gets mad at me and yells, "Leave me alone. I just don't want to talk about it." He's very angry and teary. I feel very touched by his pain. I try to be honest and share that it's been years since I've been in [a] relationship. No one wants me. Maybe I'm ugly or too hard to live with, or maybe it's the disability. I don't know, but it's lonely. I empathize with his feelings right now. We have a heart-to-heart talk. It is intense and I'm touched by his powerful pain and loneliness.*[103]

Barb's dream reflects her deep feelings for her brother. Despite his reckless driving, she put aside her anger when she saw him in need. Her dream shows her ability to place herself in her brother's shoes and feel his pain.

A woman in her sixties dreamed of being compassionate toward her ex-husband:

> *I dreamed that I had true empathy for my first husband and the loss of his girlfriend of 15 years. He is not a man who tells the whole story and keeps so much to himself. I dreamt that I kept trying to help him through this horrific time.*[24]

Healthy schemas support profound interpersonal connections. When our schemas process dreams of empathy well, we experience the emotions of others as though they were our own, which fosters a deep connection. When processed poorly, we empathize with a group or another person, but this connection makes us uncomfortable, perhaps triggering anxiety.

Our capacity for compassion may grow through recurring dreams of empathy as emotional selection works to improve related schemas. If you often dream of empathy and these dreams are accompanied by anxiety or other unpleasant emotions, you can reflect on your perceptions. You may recognize you harbor conflicting emotions that keep people at bay, or may have suffered a recent misfortune that makes it difficult for you to be happy for others. Regardless of your perceptions, reviewing those that limit your ability to be empathetic is a step toward enhancing related schemas. As these schemas improve, emotional selection can build on your progress to further develop your ability to empathize with others.

Other Belongingness and Love Themes

The schemas associated with social interactions are complex. The challenges associated with relationships between parents, friends, family, lovers, and acquaintances fill countless psychology and self-help books.

The above sketch on belongingness and love merely scratches the surface of this complex topic. I could have certainly explored additional needs, such as sympathy. Rather than comprehensively describing every identifiable need associated with belongingness and love, I show how deconstructing such dreams while awake can be a step toward improving the underlying schemas.

Esteem Dreams

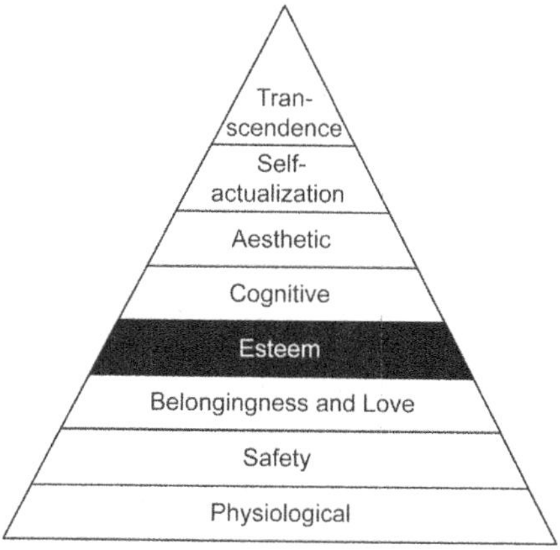

Studies have shown that people with high self-esteem are more resistant to depression when faced with trauma, stress, and misfortune,[104] and more capable of coping with difficulty in general.[105] Such findings explain why emotional selection may work to elevate our self-esteem while we sleep.

Self-esteem stems from an appreciation of our mastery, adequacy, efficacy, independence, relationships, status, and other facets of our character. So, our dreams often incorporate pleasant themes in which we demonstrate these qualities, and unpleasant themes in which our need to possess these qualities is unmet.[106]

Status Themes: Embarrassment Versus Prestige

Elevated social status helps us navigate work, family, school, relationships, and other aspects of our life. Securing a respectable job or receiving a coveted award can be self-affirming, which can contribute to success in other areas. Our friendships and close relationships likewise make us feel good about others and ourselves.

Schemas must cope during the unfortunate times when our status takes a hit and the good times when our status is elevated. So, we may regularly have unpleasant dreams of embarrassment, and pleasant, polar opposite dreams of prestige.

Thwarted Need for Status: Dreaming of Embarrassment

In this unsettling dream, we find ourselves in an embarrassing situation. We may have forgotten to dress and be accidentally naked in public, or need to use a washroom but cannot find the requisite privacy. Our physical attractiveness may lessen, as we suddenly go bald, our teeth fall out, or we suddenly age. In the dream, people may not notice our predicament, but we feel socially vulnerable, nonetheless. Regardless of the dream scenario, we are in the company of others and feel embarrassed.[107] (Dreams that trigger feelings related to mortality rather than vanity are described under Thwarted Need for Survival: Dreaming of Imminent Death, page 129.)

A man in his thirties dreamed:

> *I was at work, sitting in my cube naked. I was embarrassed to be naked but wasn't really trying to cover up as I was talking to passersby.*[24]

People with well-developed self-respect treat themselves and others respectfully. They practice forgiveness and embody the understanding that no one is perfect and accidents happen. The man in the above dream felt embarrassed, but appeared to accept his predicament as he chatted with a passerby.

A woman in her fifties had the common dream of teeth falling out:

> *I dreamt that I was on holiday with my fairly new partner and two of my teeth fell out while I was brushing them. I*

panicked about what he would think if he saw me with gaps in my teeth and in my nervous state one of them fell down the plug hole. The other I managed to secure back in my mouth. At this stage, I woke up feeling very anxious but soon felt the relief of realising it was only a dream.[24]

Dreams of embarrassment are never enjoyable. When schemas process these dreams well, we forgive ourselves, understanding that everyone makes mistakes, and that our situation does not define us. When processed poorly, we see ourselves as deeply flawed and deserving of ridicule.

The overwhelming feelings that often accompany recurring dreams of humiliation may progress toward self-forgiveness as emotional selection adapts schemas toward coping with an unmet need for status. When these dreams show no progression, you can assess related perceptions while awake. You may find that you are overly sensitive to challenges to your perceived status; that you are self-critical, feeling ashamed by your social mistakes; or perhaps your need for acceptance from others is too great, which contributes to your fear of public humiliation. Regardless of your perceptions, evaluating those that foster embarrassment while awake is a step toward enhancing related schemas. As your schemas improve, emotional selection can build on your progress while you sleep to further adapt your ability to cope with humiliation.

Satisfied Need for Status: Dreaming of Prestige

The antithesis of dreaming of embarrassment is dreaming of prestige. In these enjoyable dreams, our status is elevated. We may captain a ship, pilot an airplane, or hobnob with influential people, such as world leaders or celebrities. Variations include an enjoyable sexual encounter with a famous person. Regardless of the dream scenario, we are excited and at ease.[108]

A law student dreamed of an important paper she was writing during her waking life:

> *I had been appointed to the Supreme Court. I was sitting there talking to Thurgood Marshall. Suddenly I remembered I had a paper to write and jumped up. Then I realized I didn't have to worry about it anymore; I already was on the Supreme Court. I sat down and happily continued the conversation.*[109]

The dreamer felt at ease in a high-profile position, indicating her schemas coped effectively with prestige. In her dream, emotional selection additionally tested her schemas by introducing the stressful scenario in which she forgot to write a paper. Her schemas continued to perform admirably, as she embraced her newfound status despite her forgetfulness.

A man in his twenties dreamed of winning a prestigious award:

> *I saw a stage, and I was on it. I was performing Dr. Caius, and the audience was one of the best I've ever performed for. Then the president came up and gave me an outstanding achievement in arts [award].*[24]

Confidence and self-respect are often products of our environment. Adults who were treated respectfully as children have a greater likelihood of fostering self-respect as adults. Those not raised in a mutually respectful environment may struggle with respecting themselves and others.

When our schemas process enjoyable dreams of prestige well, we feel worthy of our newfound status. When processed poorly, we may be excited initially, but low-grade anxiety eventually comes to the fore. We may chat with a historical figure or celebrity who is fully engaged, yet question our ability to hold his or her interest.

Recurring dreams of prestige may grow more enjoyable over time, as emotional selection adapts schemas to meet this important need. If you often dream of prestige and are unable to embrace important dream moments, you can review related perceptions while awake. Perhaps a parent, caregiver, or influential person in your life feels threatened when you gain prestige, or your personality naturally shies away from the spotlight. Regardless of your perceptions, identifying those that hinder your ability to enjoy an elevated level of prestige is a step toward enhancing related schemas. As these schemas improve, emotional selection can build on your progress while you sleep to further develop your ability to embrace prestige.

Mastery Themes: Poor Performance Versus Excellent Performance

Demonstrating exemplary performance can contribute to a sense of well-being, particularly when we consider our skills as important and our contribution as significant. On the flip side of the performance coin are those occasions in which we perform poorly, which can challenge our sense of self.

The belief in one's ability to perform well can manifest as self-fulfilling prophecies. Pleasant dreams of excellent performance test whether our schemas support a self-image that embraces occasions in which we demonstrate excellence. Their polar opposite are unpleasant dreams, which test our ability to cope with occasions when our performance is lacking.

Thwarted Need for Mastery: Dreaming of Poor Performance

In this unsettling dream, we perform poorly while fulfilling an important role. We may be at work and bungling our project, at home and our parenting is subpar, or in school and flubbing an exam. Typically, others in the dream perform adequately, and only our performance lacks. (Similar dreams in which the problems are not a result of our performance, such as unfair testing conditions, are described in Thwarted Need for Efficacy: Dreaming of Powerlessness, page 185. Dreams of ineptitude are described in Thwarted Need for Adequacy: Dreaming of Ineptitude, page 181.)

A woman in her twenties dreamed:

> *[I] did something really stupid at work (I can't remember what) and was positive I was going to get fired. When I woke up, [I] was sure I was without a job.*[24]

In dreams of performing poorly, an authoritative figure, such as a boss or professor, may observe our performance. Alternatively, we may scramble to conceal our inadequate work from others. Regardless of the scenario, our poor performance is the focal point of the dream.

A man in his fifties dreamed:

> *I was having difficulty at work with whatever I was doing. I felt as though I had botched it, and my boss was basically telling me that I had "lost it"—whatever it was that had prompted him to hire me was no longer there. Acknowledging that I had been wildly successful until recently, he felt compelled to fire me. This was especially disturbing because we had been friends for many years.*[24]

Confidence under pressure calms nerves and improves focus. When schemas process dreams of poor performance effectively, we perceive ourselves as capable and remain calm. We accept that everyone makes mistakes, view our shoddy performance as a life lesson, and forgive ourselves. When processed poorly, we perceive ourselves as somewhat flawed, unable to master important tasks.

The disappointment experienced during dreams of poor performance may lessen over the course of recurring dreams. If you often dream of poor performance, and your ability to cope during these dreams is not improving, you can review related perceptions while awake. Perhaps you find that you are too self-demanding, that you have taken on too many roles, or believe your mistakes result from deeper flaws. Regardless of your perceptions, as you reflect and realize that even great masters occasionally misstep, emotional selection can build on your progress while you sleep to further improve your ability to cope with moments of poor performance.

Satisfied Need for Mastery: Dreaming of Excellent Performance

In this gratifying dream, we perform admirably. We may find ourselves taking an exam and the answers come easily, or performing a difficult task with a deft touch. Variations include dreams of excelling in a play or sporting event. Afterward, people may congratulate or thank us for our performance. (Dreams of our ability to complete routine tasks are described in Satisfied Need for Adequacy: Dreaming of Competence, page 183.)

A man in his fifties dreamed of excellent performance at work:

> *I was working in a mine. Was around others I seemed to know or had previously worked with. They were familiar. Awoke with the thoughts of being honored to do the specialized work I had participated in. Probably metal work, though I have never worked in mines.*[24]

A woman in her forties also dreamed of demonstrating mastery at work:

> *I am standing in front of a group of clients giving a presentation. I end the presentation and ask for questions. Questions are answered and clients sign up. My boss and the board of directors shake my hand and tell me well done.*[24]

Dreams of excellent performance often begin with a mix of anxiety and excitement, as the dream scenario may involve an important or difficult task. After demonstrating competence, any nervousness we experience fades as we execute flawlessly. When well-adapted schemas process dreams of excellent performance, we bask in our success, feeling worthy of recognition and appreciation. When processed poorly, anxiety ensues. Despite continued excellent performance in the dream, we dismiss each accomplishment as a fluke and brace for a perceived inevitable failure.

The satisfaction that often accompanies dreams of excellent performance may grow over time as emotional selection adapts schemas toward meeting this important need. If you often experience these dreams without progression, you can investigate related perceptions while awake. You may realize that your parent, guardian, or friend feels threatened or jealous when you demonstrate mastery that he or she lacks, or that your personality may support others in their pursuits, but does not accommodate your success. Regardless of your perceptions, identifying those that inhibit your ability to enjoy moments of mastery is a step toward enhancing related schemas. As your schemas improve, emotional selection can build on your progress while you sleep to further develop your ability to embrace excellent performance.

Adequacy Themes: Ineptitude Versus Competence

Competency is important. Missing a bus to a job interview could lead to a lost employment opportunity, resulting in hardship. Although most situations do not have such exponential ramifications, life's minor slips can carry consequences.

It is the rare person who is truly inept, yet many of us wonder if we are that person. We may have imposter syndrome, feeling inferior to our peers, or we may feel anxious in settings in which everyone else seems so calm.

Like other dream categories, adequacy dreams have pleasant and unpleasant themes. Unpleasant adequacy dreams, in which our need for adequacy is thwarted, test our ability to cope with times when our competence comes into question, and pleasant adequacy dreams, in which our need for adequacy is satisfied, ensure our schemas include a self-image that supports competence.

Thwarted Need for Adequacy: Dreaming of Ineptitude

In this trying dream, routine tasks befuddle us. We may rush to catch a bus, plane, or train, but become confused and miss it.[110] We may attempt to host a dinner party and cannot find the ingredients or cookware to make the meal, or may be at work and struggle to perform routine tasks. We may find ourselves back in school and taking an exam we neglected to study for, in a theatrical performance in which we forgot our lines, or in a sporting competition for which we did not study the team playbook. Typically, everyone in the dream is prepared except us, and we feel there is no one to blame but ourselves. (Dreams in which the problems are not a result of our being unprepared, such as testing conditions that are unfair, are described in Thwarted Need for Efficacy: Dreaming of Powerlessness, page 185. Dreams in which malfunctioning equipment or chaos is to blame are described in Thwarted Need for Stability: Dreaming of Chaos, page 150. For dreams in which our knowledge is lacking, see Thwarted Need for Knowledge: Dreaming of Ignorance, page 197.)

A woman in her forties dreamed:

> *I am late for a test. I knew there was a test, but I didn't know I was enrolled. [I'm] fuzzy between [whether] this is now or back in college. I'm not ready for the test at all, but I still move like I know what I'm doing. I enter the room [to] take the exam, sit down, and stare at it. I am scared on the inside, but I struggle to remember anything, anything I can possibly think of to fake my way through. Maybe I can pass, but I have no idea how.*[24]

Another woman in her forties had the common recurring dream of finding herself on stage and unprepared:

> *I'm about to take the stage but haven't learned my lines. This [dream was] slightly worse, as I couldn't even make sense of my costume.*[24]

Anxiety and frustration often accompany dreams of ineptitude. When our schemas process these dreams well, we keep events in perspective, realizing that even competent people make mistakes or that our failure is a minor setback. Consequently, missing a bus or failing to operate technology during the dream feels like a minor inconvenience, rather than being unbearable. When processed poorly, we feel discouraged, seeing our failure as affirming our ineptitude.

The disappointment that often accompanies dreams of an unmet need for adequacy may subside over time as emotional selection adapts related schemas. If you frequently dream of being inept, you can review your perceptions that challenge your sense of competence. You may recognize that you recently had some bad luck that reinforced your negative perceptions, or that the jobs or roles you perform are simply outside of your strengths. Alternatively, perhaps important people in your life highlight your failures and ignore your successes, or you are overly self-critical. Regardless of your perceptions, reconsidering those that contribute to a diminished sense of adequacy can lead to a greater sense of competence. As your schemas improve during waking, emotional selection can build on this progress while you sleep to further develop your ability to cope with the occasional stumbles that everyone experiences.

Satisfied Need for Adequacy: Dreaming of Competence

In this enjoyable dream, we have the magic touch. We may be traveling, attending an event, or spending an evening out. Whatever the situation, we complete tasks and meet everyone's expectations swimmingly. We may fix something that broke, such as an automobile or piece of machinery. Initially, we may be unsure of what to do, but find that we consistently guess correctly.[111] (Dreams in which we excel in an important event, such as an exam, theatrical performance, or sporting event, are discussed in Satisfied Need for Mastery: Dreaming of Excellent Performance, page 179.)

A man in his sixties dreamed of demonstrating a routine skill in front of varying friends:

> *[I was] in car with a friend, changing to different friends (current and past) while driving over a bridge with water [nearby] and then far below. Then I am driving the car, which surprises me because I do not drive and never had a driver's license. I feel comfortable driving and wonder why I do not drive; it's so easy.*[24]

A man in his twenties dreamed of competence at work:

> *Working in a metal shop similar to a past job in real life. The difference is I had just started and was being shown around the place and told to try new tasks. I remember easy tasks seeming difficult, but I simply relaxed and things became very easy.*[24]

Dreams of competence can be deeply satisfying. When processed well, we may feel initial anxiety at the occasional dream snafu, but these feelings quickly fade as we improvise perfectly. When processed poorly, things go smoothly, but our enjoyment fades into anxiety as we question our abilities. Despite demonstrating adequacy in our dream, we feel unprepared or unqualified, and are relieved when we reach our goals.

The satisfaction that often accompanies dreams of competence may deepen over the course of recurring dreams as emotional selection adapts related schemas. If you frequently dream of adequacy and find that you are unable to embrace dream moments, you can review related perceptions while awake. Perhaps your personality struggles with acknowledging your abilities, resulting in a fear of becoming immodest or self-congratulatory, or influential people in your life may be jealous of your competence. Regardless, reviewing your perceptions may lead to the realization that you perform most of your daily tasks competently. Emotional selection can build on these insights while you sleep to adapt your schemas to further embrace your competence.

Efficacy Themes: Powerlessness Versus Empowerment

Occasionally, we must influence the actions of others. Someone may be physically encroaching on our personal space, requiring us to restore an appropriate distance, or they may endanger themselves, and we must take action to protect them. Whatever the situation, our perceptions of efficacy are essential for changing the course of events in our life. So, emotional selection processes dreadful dreams of powerlessness that test our ability to cope when our need for efficacy is thwarted, and invigorating dreams of empowerment that test our ability to cope with occasions when our need for efficacy is satisfied.

Thwarted Need for Efficacy: Dreaming of Powerlessness

In this unsettling dream, others treat us unfairly. We have an opportunity to confront their behavior, but question ourselves. The longer we delay, the more anxious we become. We may be a terrified passenger in an automobile being driven recklessly, a student in a classroom with a dictatorial professor, or an employee with a bully for a boss. The scenario is often public, and the perpetrator is an authoritative figure. Obstacles in the dream may prevent us from performing a task that will be scrutinized by the authoritative figure, such as an examination with indecipherable words or pencils that repeatedly break. The authoritative figure may be intentionally unfair or oblivious to the problem. Regardless of the dream scenario, we hesitate to assert ourselves for fear of creating a conflict. (Dreams of difficulty because of a lack of preparedness are described by Thwarted Need for Adequacy: Dreaming of Ineptitude, page 181. Dreams of a violent attack or sexual assault are discussed in Thwarted Need for Protection: Dreaming of Vulnerability, page 146.)

A woman in her fifties dreamed of being asked to work beyond her job description:

> *Employers from thirty years ago expected me to do work that I wasn't qualified for and didn't know how to do. I didn't speak up to get myself out of it.*[24]

During dreams of powerlessness, we perceive that we are being exploited, which makes us anxious, insecure, or angry. When our schemas process dreams of powerlessness well, we act, overcoming our anxiety and expressing ourselves in a calm, non-aggressive manner. When processed poorly, we consider ourselves ineffectual and fail to act, or act inappropriately by lashing out.

A dream from my personal dream log placed me back in high school, where I took an examination under unusual conditions:

> *I was sitting in the back seat of a moving automobile while the instructor read the exam questions from the passenger seat as the driver made sharp turns, making the answers to the exam difficult to write. I felt like I could have done OK on the exam if the conditions didn't make it so hard to take. I considered asking the driver to slow down but was too intimidated.*

Dreams of powerlessness often feel overwhelming. When schemas process these dreams effectively, we feel powerless, but understand that we are in a difficult situation that other people likewise find challenging, or we may feel emboldened and rectify an unfair situation in our dream by working with authoritative figures. As we do, peers may support our opinion and express relief. When our schemas process these dreams poorly, we may perceive ourselves as ineffectual, feeling that others would have easily rectified the issue at hand.

The discomfort that often accompanies dreams of powerlessness may ease over the course of recurring dreams as emotional selection adapts related schemas. If you often dream of feeling powerless without this dream progression, you can review related perceptions while awake. You may find that you delegate too much authority to others, or that people in your past demanded an inappropriate amount of control over your decisions. Regardless of your perceptions, identifying those that limit your efficacy is a step toward repairing related schemas. Emotional selection can build on this progress while you sleep to further improve your ability to navigate situations that require efficacy.

**Satisfied Need for Efficacy:
Dreaming of Empowerment**

In this rewarding dream, we find ourselves able to influence those around us. We may motivate others toward a common goal of rescuing injured or abused people or animals. Regardless of the dream scenario, we take control of a difficult situation and correct it. Although dreams of empowerment are rarely pleasant in the conventional sense, they can be satisfying when we remedy a problem.[112]

Helen was unhappily married to Dick, a self-indulgent man who was having an affair. After Helen ended their relationship, she went through a period of depression and subsequently rebounded. She attended law school and found a boyfriend who was more mature than her ex-husband. During this period, she dreamed of empowerment:

> *I was on a trip with about 30 people, an adventure trip in Japan. We were going to climb a mountain. We were in a valley with very steep walls and it was very, very foggy and no one could see. Everyone was wandering around with their arms out in front of them, like they were sleepwalking and trying to figure out where to go. Finally, I said, "This is ridiculous, I'm going to find my way out of here." So, I put my arms down and opened my eyes. I found a rock that went up like a stairway. And everyone started to argue about who should go first. I thought I should go first because I found it. So, I started to walk up the stairs going up the side of the mountain, and everyone started to follow me, walking up the stairs single file slowly. I remember feeling very anxious and afraid at first, but when I found the staircase, I felt very,*

> *very good, real[ly] strong. I felt I was the only one [who] could do it. I had to do it myself. So, I did.*[113]

Michael, whose dream of passively escaping a flood we reviewed earlier, dreamed ten months later that he actively averted a similar catastrophe:

> *I am returning to my house in a huge, wooded area with other houses on large lots. As I approach, I see hundreds of backhoes tearing down the forest around my house. I can hear their engines throttling. It reminds me of pictures of the destruction of the Amazon rainforest. I run up to the foreman and shout at him, "You are destroying my house and my whole neighborhood." I was outraged and started ringing all of my neighbors' doors, urging them to join me in fighting back. I organized a protest group to sit in front of the bulldozer. I was pissed off to the point of violence. The police came to break it up, but it became violent, and they began chasing and beating people. Finally, the cops agreed to stop the destruction of trees and houses. I felt satisfied.*[114]

In Michael's dream, emotional selection tested his ability to confront injustice and influence others. Rather than succumbing to the dream scenario, as he did in his flood dream, he affected positive change, demonstrating that his schemas improved. Although his dream began with a stressful scenario, his ability to effect change resulted in a satisfying dream.

When our schemas process dreams of empowerment well, we take control of a difficult situation. We may initially feel anxious when assuming authority, but eventually feel confident in our decisions. When processed poorly, we may perceive ourselves as unworthy of our newfound empowerment, and consequently feel overwhelmed or defensively lash out at others.

The satisfaction that often accompanies dreams of meeting our need for efficacy may grow through recurring dreams. If you frequently dream of empowerment and struggle with leading others during these dreams, you can review related perceptions while awake. Perhaps you had a controlling parent or caregiver who damaged your sense of efficacy, or you prefer to avoid the confrontations that can accompany leadership. Regardless of your perceptions, reviewing those that inhibit your ability to assume power is a step toward enhancing related schemas. As your schemas improve while awake, emotional selection can build on your progress while you sleep to further develop your ability to act with efficacy.

Independence Themes: Subjugation Versus Autonomy

As social beings, we routinely depend on others. However, to successfully fulfill our needs, we must also be capable of acting independently, which often involves striking a balance between our need for independence and our obligation to obey the rules of society. Typically, we adopt a moral framework that incorporates accepted social norms and an internal ethical code. When making decisions, our mind continually references schemas that model this framework, ideally forming a perception with minimal influence from authorities.

Our schemas must cope with times when our need for independence is both met and unmet. Unpleasant dreams of subjugation test our ability to handle occasions when we lose our independence, while pleasant dreams of autonomy test our schemas' ability to embrace times when we act independently.

Thwarted Need for Independence: Dreaming of Subjugation

In this distressing dream, authoritative figures jeopardize our independence. Teachers, parents, or police scrutinize our behavior because of something unethical or illegal we did, or because we were in the wrong place at the wrong time.

Marcia, a married woman who relayed that she had difficult experiences with men, dreamed:

> *I was in the supermarket and saw a boy about ten years old with his mother and little sister. Every time the mother turned away the boy would pinch his sister. When she cried, the mother would tell her angrily to hush, to behave, while the boy stood and grinned behind their backs. Then in the check-out line they were right behind me when a guard came up and said we all had to be searched for shoplifting. The guard opened my purse and pulled out an expensive can of caviar. He put handcuffs on me and all the time I was trying to tell him that nasty boy had framed me. It was awful and I woke up trembling.*[115]

Marcia was a bystander in her dream. She professed her innocence, which is a good sign that her schemas coped with being accused wrongly. However, her trembling suggests that the events of her dream affected her deeply.

A woman in her forties dreamed of elaborate mischief that led to her subjugation by police:

> *I dreamed that I was going around hopping into semi-trucks that were left with [the] motor running and I would drive them. Well, when I got into about the third or fourth truck, I couldn't manipulate the turns and I wrecked the truck into some fences and houses. It was nighttime and everyone came out of their houses to see what the commotion was. I fled and ran around, peeping to see when the cops would show up, and I said to myself, "Why am I running? I didn't do*

anything wrong; it was just an accident and I wasn't drunk." So. when I went to the scene, I confessed that I had run for "help" and came back because I didn't want them to think that I was doing a hit-and-run. But when they started questioning me about drinking, I said, "Absolutely not," and the cops kept saying, "Come on now, you better tell us the truth or else take this breathalyzer test." So, I said, "Yes, give me the test to prove it." Well, when I blew into it, it came up as having alcohol content in my system, and I screamed, "No, no, no! I know I didn't drink!" So, I said, "Do it again now!" I did and it came back with a higher content! The police woman was laughing at me and shaking her head saying, "Stop denying it. You got caught!" I then found my mom waiting for me to see what was going on, and when I told her what happened, she shook her head in disappointment. I said, "No one believes me!" After I swore I didn't drink, she believed me and said, "Get in the semi, we're going to get the heck out of here!" So, we get in [and] drove off very, very fast on a dirt road, leaving all this dust behind, and she said, "This is how you drive this thing!" All I could think about was how much trouble I was going to be in, going to jail. Such a nightmare! I woke up feeling like I was really in trouble, but slowly realized it was a dream. All I said to myself was that I did not ever want to drink and drive.[24]

What one group considers mischievous or immoral, another group may regard as acceptable or desirable. Consequently, when our schemas process dreams of subjugation effectively, we believe in ourselves despite the judgment of others. When processed ineffectively, our conflicts with authorities cause us to question our moral compass and our self-worth.

The stress that often accompanies dreams of subjugation may diminish over the course of recurring dreams as emotional selection adapts schemas toward coping with an unmet need for independence. If you frequently dream of subjugation without this progression, you can help emotional selection by assessing related perceptions while awake. You may find your self-image is influenced by powerful figures in your life, such as a parent or domineering friend; that religious ideals or social norms do not align with your values; or that you have a fear of being controlled by others. Regardless of your perceptions, reconsidering those that struggle with subjugation is a step toward improving the underlying schemas. As your understanding evolves and your inner values emerge, emotional selection can build on your progress while you sleep to improve your ability to cope with subjugation.

Satisfied Need for Independence: Dreaming of Autonomy

In this rewarding dream, we exercise our independence. Others may try to sway us, but we remain steadfast. We may find ourselves before the police or other authoritative figures, but believe we have done little or nothing wrong and confront them without hesitation. We perceive ourselves as a worthy person and do not fear authority.

I once dreamed:

> *I was an adult walking along a snow- and ice-covered road in my childhood neighborhood when a friend and I noticed what looked like a pothole. Upon examination, we found it was an opening into a cavern. In the dream, storm water had completely washed away the road but left the ice behind, creating a huge network of interconnected ravines covered with an enormous sheet of ice. My friend was afraid to explore the new discovery, but I thought it would be adventur-*

> *ous, so I went on alone. The ground was now twelve feet below where the road once was, and the translucent sheet of ice above acted as an enormous skylight. Long, soft icicles hung from the ice and scattered the light, creating a beautiful reflective world. As I explored the ravines, I could see the undersides of the cars driving on the road above, their headlights shining through the snowy ice making the experience even more fantastic and beautiful. Suddenly, it occurred to me that the heavy cars could crash through the ice and injure me, so I headed back toward the surface. As I emerged, I found that my friend had gotten nervous about my adventure and called the police. I was irritated by this and slightly intimidated by the situation, but felt I had done nothing wrong. Once I began speaking with the officers, I relaxed. They were cordial and receptive, eventually letting me go.*

Like many dreams, my dream encompassed multiple themes, including discovering new places, as described in Satisfied Need for Knowledge: Dreaming of Learning, page 198. Toward the end of the dream, when I met with the authoritative figures, I was initially anxious. However, I reminded myself that I had not committed a crime, which calmed my nerves and allowed me to cope with the situation.

Dreams of autonomy do not always include themes of being innocent. A woman in her fifties dreamed:

> *I was being arrested for something I did, surrounded by police and a SWAT team. Local news was there and the FBI. The night was lit up with lights and all the neighbors were out. Funny thing is, I was very calm and not scared at all.*[24]

Dreaming of authorities signifies emotional selection is working on our schemas related to independence. When schemas process dreams of autonomy well, we are free from fear of subjugation, as any nervousness we feel fades and we navigate the situation calmly and confidently. When processed poorly, our anxiety persists in the face of authority.

The satisfaction that often accompanies dreams of autonomy may grow through recurring dreams as emotional selection works to satisfy this important need. If you often have these dreams without this progression, you can review related schemas. You may find you allow people in positions of authority to sway your opinions, or that powerful people intimidate you. Regardless of your perceptions, reviewing those that inhibit your sense of autonomy while awake is a step toward enhancing related schemas. As your schemas improve, emotional selection can build on your progress while you sleep to help you enjoy more autonomy in your life.

Recognition, Privacy, and Other Esteem Themes

Esteem is a complex topic. My sketch above, which explores the relationship between emotional selection and esteem needs, is far from comprehensive. I could have included an investigation of recognition, privacy, and other subcategories of esteem needs. Instead, I selected common examples to show how dream deconstruction applies to virtually any subcategory of our need for esteem.

Like other need subcategories, we can further refine our esteem needs. For instance, our need for independence can be subcategorized as intellectual, intuitive, emotional, or physical independence. Regardless of the granularity in which we choose to explore our needs, the process of dream deconstruction remains the same.

Cognitive Dreams

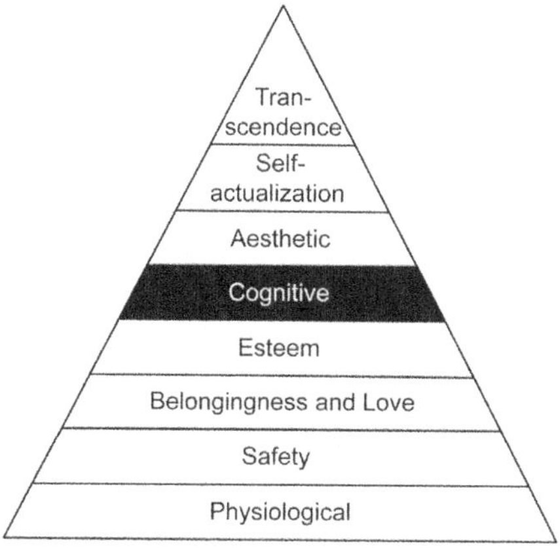

Cognition encompasses creativity, knowledge, and problem-solving. Although it might seem like cognition is what makes us human, animals also routinely exhibit cognitive needs. Monkeys have been observed solving puzzles even when they expected no reward, leading researchers to conclude that monkeys engage in puzzle-solving purely to learn.[116] Similarly, Maslow observed that some piglets wandered away from their mothers, seemingly to explore and learn. When their pen door was opened, the piglets moseyed toward the opening and eventually ventured out. When the door was closed, the piglets squealed with anxiety at being separated from their mother. However, returned to their pen, they eventually headed toward the open door again, apparently wanting to resume their exploration.

Cognition provides obvious evolutionary advantages. An animal curious to know what lies beneath rocks may find additional food sources. Another who is curious about what lies beyond a hill may find new shelter. In short, cognition helps animals and humans meet other needs.

Since nobody possesses limitless cognitive abilities, we must manage the inevitable gaps in our cognition and appreciate the instances when our cognitive abilities serve us well. So, we dream of times in which we meet our cognitive needs, and times in which our cognitive needs are unmet.

Knowledge Themes: Ignorance Versus Understanding

Throughout our life, we lack knowledge. As children, caregivers routinely place food before us with textures and tastes that we do not know. In school, teachers present a stream of new-to-us information. In our romantic relationships, we continuously find ourselves in situations in which our knowledge is limited, ranging from our first kiss to raising children.

The counterpart of ignorance is understanding. As we gain knowledge, we often share it with others. A classmate who missed a lesson may turn to us for help, a newly hired coworker may need some instruction on our company's systems, or a passerby may simply ask for directions.

Whether we are gaining or sharing knowledge, our schemas guide our behavior. When ignorant, does our limitation trigger defense mechanisms or curiosity? If our answer is curiosity, do we proceed with trepidation or ravishment? After acquiring knowledge, do we hoard it to maintain a position of power or share our knowledge to help others? If the latter, is our goal to help the mentee or merely demonstrate our knowledge? Our schemas possess the answers to these questions.

As with other need categories, knowledge dreams can be pleasant or unpleasant. Unpleasant dreams of knowledge test whether our schemas adequately cope with instances when our knowledge is lacking, and their pleasant counterparts test whether our schemas can cope with times when we are knowledgeable.

Thwarted Need for Knowledge: Dreaming of Ignorance

In this unsettling dream, our knowledge is lacking. We may find ourselves bewildered at work or school, lost in a sea of information. Variations include being lost in the literal sense, wandering an urban or natural environment, or searching for something we misplaced, such as our parked car or our wallet. Regardless of the dream scenario, we feel foolish and dumbfounded. (For dreams of losing something valuable, see Thwarted Need for Financial Security: Dreaming of Lost Property, page 142. Dreams of being simultaneously lost and chased or attacked, or in other ways feeling physically vulnerable, may fall under Thwarted Need for Protection: Dreaming of Vulnerability, page 146. Poor performance because of a lack of preparedness is described in Thwarted Need for Mastery: Dreaming of Poor Performance, page 177.)

A woman in her fifties had the common dream of being lost in high school:

> *[I was] lost in a high school corridor. Lots of people. Can't find my way. Have no idea what class I am supposed to be in.*[24]

A well-educated woman in her twenties confessed she had "a fear of being stupid." She also dreamed of being back in high school, where her fear came true:

> *In my school, I was having some major reading problems, and I didn't understand anything.*[24]

In dreams of ignorance, we lack the knowledge to perform a task or solve a problem. This makes us feel vulnerable. When our schemas process these dreams effectively, we see the events as a temporary setback that could happen to anybody. When processed poorly, anxiety consumes us.

The stress that frequently accompanies dreams of ignorance may diminish over time as emotional selection adapts schemas to meet our need for knowledge. If you frequently dream of knowledge without this progression, you can review related perceptions while awake. You may recognize that a parent or guardian was an intellectual bully who used knowledge to make others feel inferior, or that you question whether you are knowledgeable enough to satisfy the roles you have assumed. Regardless of your perceptions, reviewing those related to your perceived knowledge is a step toward enhancing underlying schemas. As your schemas improve, emotional selection can further help you cope when your need for knowledge is unmet.

Satisfied Need for Knowledge: Dreaming of Learning

In this gratifying dream, we discover something new and important. We may find a secret door that leads to undiscovered rooms in our home, gain insights from books, or uncover an ancient artifact that improves human understanding. Variations include discoveries in nature, including caves, fields, or undiscovered settings. (Dreams of discovering valuables that result in financial gain are described in Satisfied Need for Financial Security: Dreaming of Acquired Property, page 143.)

A woman in her thirties dreamed of learning with her friends:

> *I was at school and having fun with my friends, while learning and creating new things and [agendas]. I awoke with a smile.*[24]

Feelings of joy and satisfaction often accompany newly acquired knowledge. Scientist and author Carl Sagan dreamed of being privy to newfound knowledge:

> *I once had a dream that will tantalize me forever. I dreamt I was idly thumbing through a thick history text. I could tell from the illustrations that the work was moving slowly, in the usual manner of such textbooks, through the centuries: classical times, Middle Ages, Renaissance, and so on, gradually approaching the modern era. But then there was World War II with about two hundred pages left. With mounting excitement, I worked my way more deeply into the work until I was sure that I had passed my own time. It was a history book that included the future.*[117]

In dreams of learning, we make an exciting discovery or solve a difficult problem. When our schemas process these dreams well, the experience is euphoric, as our world literally or figuratively expands before us. Sagan's dream tested his ability to cope with acquiring meaningful information. He relates his excitement at his discovery, indicating that his schemas performed admirably. When schemas process dreams of learning poorly, our initial excitement at our discovery fades as we become anxious, perhaps perceiving ourselves as undeserving of such important understanding.

The exhilaration that accompanies dreams of acquiring knowledge may grow as emotional selection works to incorporate this important need into our schema network. If you frequently dream of knowledge and struggle to embrace dream moments, you can help emotional selection by assessing related perceptions while awake. You may realize that people in your life feel threatened by your newfound understanding, or that you had a parent or guardian who undermined children's expanding knowledge to reinforce their position of authority. Regardless of your perceptions, identifying those that limit your ability to cope with new understanding is a step toward enhancing underlying schemas. As these schemas improve, emotional selection can build on your progress to further develop your ability to embrace knowledge.

Intuition, Creativity, and Other Cognitive Themes

Whether we are at our workplace, home, or social event, we routinely exchange ideas with workmates, family, and friends. After acquiring knowledge, we apply it to perform our job, do household chores, or help others. When not learning, we can often be found developing creative new ways to use our existing knowledge. Simply put, our cognitive needs keep us learning and mentally growing.

Like other need categories, we can further divide the cognitive needs that appear in dreams. For instance, knowledge can be divided into subcategories that include social knowledge and procedural knowledge. We also dream of other cognition needs, which I did not deconstruct above, such as our needs for intuition and creativity. Similar to virtually any dream theme, cognitive dreams of intuition or creativity can be deconstructed by identifying the perceptions and schemas in our dream and reviewing how they apply to our life. As we deconstruct our dream, we are taking a step toward enhancing related schemas. As these schemas improve, emotional selection can build on our progress while we sleep to further develop our ability to meet all of our cognitive needs.

Aesthetic Dreams

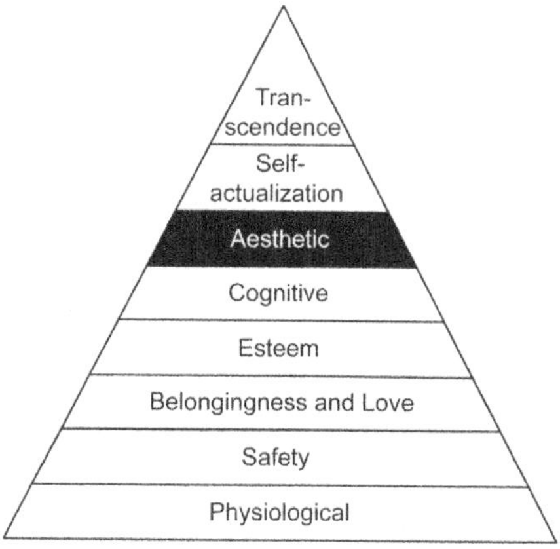

As mentioned, our abilities to meet lower needs, such as physiological and safety needs, are clearly evolutionary, but as we travel up Maslow's hierarchy, each need requires more reflection to understand its relationship to our evolutionary success. Regarding our need for aesthetics, we strive to surround ourselves with beauty, but how do such drives increase our ability to survive?

The answer may lie in the fact that experiencing beauty has measurable health benefits. A study asked a group of men and women to view paintings while a small laser zapped one of their fingers, producing pain levels similar to a pinprick. Participants rated the pain level one-third lower while viewing the paintings. Electrodes monitoring the pain centers of their brains agreed with their verbal responses.[118] In another study, a group of pregnant women listened to soothing music for thirty minutes a day while resting, doing chores, or going to bed, while another group of pregnant women did not listen to music. Two weeks later, the researchers found the music-listening group had less anxiety and stress than the control group.[119] Studies such as these agree with the opinion of artist and philosopher Robert

Bateman: "Aesthetic pleasure is not to be taken lightly; it's very important for the spirit and therefore the health."

Visual Beauty Themes: Ugliness Versus Beauty

We often go to great lengths to surround ourselves with beauty, including painting rooms in our home with pleasant colors, hanging artwork on our walls, and spending our hard-earned money on stylish clothing. When not decorating our homes or ourselves, we can often be found watching a sunset or marveling at the beauty of nature.

Like other needs, we must cope when our need for beauty is both met and unmet. So, it comes as no surprise that we dream of meeting our need for beauty and its polar opposite, our need to cope with ugliness.

Thwarted Need for Visual Beauty: Dreaming of Ugliness

In this distasteful dream, we are subjected to ugliness. We do not feel threatened by our surroundings. Rather, they depress or disgust us. We may find ourselves in a garbage dump, wasteland, or dilapidated city. We may be surrounded by grotesque creatures who mean us no harm, but they repulse us, nonetheless. Regardless of the predicament, we wish to distance ourselves from our surroundings.[120] (Dreams of feeling physically threatened are described in Thwarted Need for Protection: Dreaming of Vulnerability, page 146.)

A woman in her forties dreamed of dilapidation:

> *I am a teacher at a large one-story school on a mountain. Students vandalize the best bathroom on campus, it is large, roomy, [with] high ceilings and lots of light and rounded wooden doors on the stalls. I tell them it must be closed for*

repairs. The girl students are sad. I know the boy students who did it. The bathroom they must use now is small, old-fashioned with ugly metal stalls, [and] darker with unpleasant yellow lighting.[24]

Barb, whom we met earlier in this chapter, dreamed of being surrounded by ugliness while she swam:

I was swimming in a lake and noticed the water level was going down fast. The lake drained out entirely, and I saw creepy white crustacean[s] and fish crawling around on the [newly] exposed surface. I thought, "Ugh. I was swimming around with those ugly things?" Some were that white that denotes never [seeing] the light of day, like maggots.[121]

Barb's dream tested her ability to cope with ugliness. Since the creatures in her dream posed no danger, emotional selection was not testing her safety needs. In her dream, she did not resign herself to being surrounded by "ugly things." Instead, she clearly felt she deserved to swim in a pristine lake and reacted appropriately to swimming among creatures that resembled maggots.

Dreams of ugliness are clearly unpleasant. When schemas process these dreams well, we feel worthy of beauty and seek to escape the ugliness that surrounds us. When processed poorly, we feel resigned to our fate of being subjected to ugliness.

The discomfort that accompanies dreams of an unmet need for beauty may diminish over time as emotional selection adapts related schemas. If you frequently dream of ugliness without this progression, you can review related perceptions while awake. Perhaps you are routinely exposed to ugliness in your daily life, or consider yourself unworthy of being surrounded by beauty. Regardless of your perceptions, reviewing your ability to cope with ugliness is a step toward adapting associated schemas. As your schemas improve, emotional selection can build on your progress while you sleep to further develop your ability to cope with ugliness.

Satisfied Need for Visual Beauty: Dreaming of Beauty

This enjoyable dream introduces us to seemingly boundless beauty. We may discover a wondrously beautiful garden, find ourselves in a cave with iridescent walls, in a forest with light streaming through the trees, or in a glistening city. As we observe our surroundings, beauty consumes us.

A man in his twenties dreamed:

> *I was riding through woods on a white horse, when I came to this opening to this crystal-clear lake. When I looked down, I saw that I was wearing the shiniest knight-plate armor. On my breastplate was a gold cross surrounded in red roses [and] vines, wrapped in blue and white flames. I was at peace. As I stood there by [the] lake looking at the beautiful valley and blue sky that surrounded me, I just stood in awe of its beauty. I awoke from my dream when I got back on my white stallion.*[24]

From my personal dream log:

> *I was walking with someone—I don't remember whom—and we entered the backyard of a neighbor from my childhood. She had an old barn. I peeked in and saw my neighbor and an attractive woman walking toward the door. I was a bit embarrassed at snooping around, but my neighbor was glad to see me. She introduced me to the woman and invited me in. The garage was dilapidated, but made a really beautiful ruin with stone walls, wood moldings, an open sky above and grassy ground below. We walked up a slope and*

out what was once the back of the garage. Toward the back of the yard, we turned around and took in the vista. Her yard was amazingly beautiful. It was landscaped, but in a manner that you never would have known, since it looked so natural. The grass was especially beautiful: long, thick, and deep green.

I marveled at the landscape my mind conjured in my dream. Had I felt undeserving of experiencing such beauty, it would indicate relevant insecurities in my waking life. Fortunately, I enjoyed the environment of the dream and woke with a sense of calm and awe.

Dreams of beauty are often wondrous. If you frequently dream of beauty and feel somewhat anxious or perhaps undeserving during these dreams, you can review related perceptions while awake. You may realize you were raised to regard beauty as superfluous, being taught that pragmatic pursuits should always take precedence, or you may believe beauty is reserved for people with higher means or talents. Regardless of your perceptions, reconsidering those that relate to beauty during waking is a step toward enhancing underlying schemas. As your schemas improve, emotional selection can build on your progress while you sleep to improve your ability to embrace beauty.

Auditory, Gustatory, and Other Aesthetic Themes

As observed by the German nineteenth-century composer Robert Schumann, "The aesthetic principle is the same in every art, only the material differs." So, our senses of sight, sound, touch, smell, and taste are all capable of capturing beauty. This explains why we surround ourselves with works of art, spend precious time and income on music, and dine not only for nutrition, but also to satisfy our palettes.[122]

An engineer seeks elegant solutions, while a poet seeks eloquent prose. Because beauty comes in many forms, it presents itself in dreams in countless ways. The techniques for deconstructing beauty in dreams are the same, regardless of the type of beauty. First, identify the emotions and feelings accompanying aesthetic dreams. Second, review underlying perceptions for parallels in our waking life. Third, reconsider perceptions that inhibit our aesthetic needs. As our schemas improve, emotional selection can further adapt our ability to satisfy our aesthetic needs.

Self-actualization Dreams

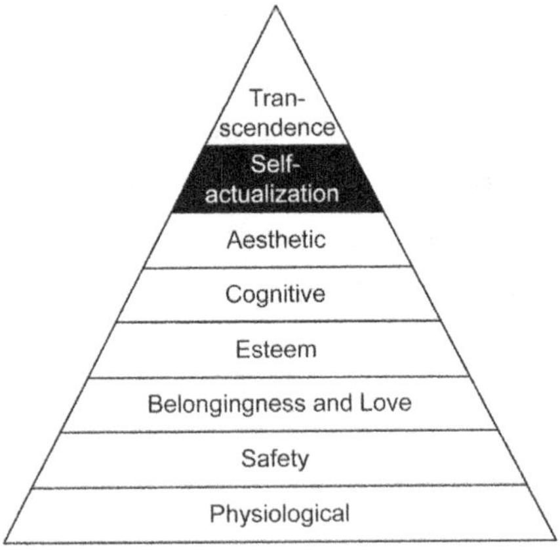

Neurologist Kurt Goldstein coined the term "self-actualization" to define an organism's drive to reach its full potential. Maslow adopted this term for the penultimate need in his hierarchy:

> It refers to the [person's] desire for self-fulfillment, namely, to the tendency for him to become actualized in what he is potentially. The specific form that these needs will take will, of course, vary greatly from per-

> son to person. In one individual it may take the form of the desire to be an ideal mother, in another it may be expressed athletically, and in still another it may be expressed in painting pictures or in inventions.[123]

Dreams can test the ability of our schemas to cope with the thwarting and satisfaction of our need to self-actualize. During dreams of thwarted self-actualization, we cannot be our best selves because of exterior or interior forces. Conversely, dreams of satisfied self-actualization can be a person's most memorable dream.

Reaching Potential Themes: Unfulfilled Potential Versus Fulfilled Potential

In an ideal world, every person's human needs for food, shelter, safety, belongingness, cognition, and aesthetics, would be met, freeing up our time to pursue our highest needs. Unfortunately, it is the rare individual who meets all but his or her highest needs. Because we typically spend most of our time pursuing lower needs, self-actualization dreams and transcendence dreams are less common. However, while relatively few dreams feature our highest needs, those that do are typically poignant and beautiful.

Thwarted Need for Reaching Potential: Dreaming of Unfulfilled Potential

In this rare and unsatisfying dream, we are blocked from reaching our full potential. We see our path to fulfillment, but people or obstacles stand between us and our destination. We feel worthy of self-actualizing, but our best efforts are unrealized due to circumstances beyond our control. Our inability to self-actualize may embarrass or frustrate us, or trigger disappointment and sadness.[124] (Dreams of poor performance that result in embarrassment are discussed in

Thwarted Need for Mastery: Dreaming of Poor Performance, page 177. Dreams of failing to meet basic expectations are described in Thwarted Need for Adequacy: Dreaming of Ineptitude, page 181.)

A woman in her fifties dreamed of unfulfilled potential from an otherworldly perspective:

> *I was watching black clouds rolling toward the west, and a bright light was shining in the east. Gold sparkles of light were being drawn into this light that kept getting brighter. I realized it was the end of the world and God was drawing all the saved souls into this light and each golden spark was the soul of a believer and he was collecting them to his spirit. I asked why I was not taken, and he said I had a purpose to fulfill and it was not my time, but the day would come.*[24]

A man in his twenties dreamed of unfulfilled potential in this world:

> *I was with an old friend I hadn't seen in years. I have the strong feeling that we need to reach a goal or purpose of some kind, but my friend keeps falling off-course, dragging me along and wasting our time on meaningless things. We never reach the climax, I never feel fulfilled.*[24]

A dream of thwarted self-actualization adapts our schemas toward realizing our potential. When our schemas process these dreams well, we understand that our inability to fulfill our potential is due to forces beyond our control and does not reflect our ability to self-actualize in the future. When processed poorly, our failure in the dream may trigger anxiety about our capability to reach our potential, or frustration at the delay of our self-actualization.

The disappointment that accompanies dreams of thwarted self-actualization may wane over time as emotional selection adapts related schemas. If you often dream of being unable to meet this high need without progression, you can review related perceptions while awake. Perhaps important people in your life would be jealous if you self-actualized, or you question your ability to reach your potential. Regardless of your perceptions, identifying and reconsidering those that limit your ability to self-actualize is a step toward enhancing underlying schemas. As these schemas improve, emotional selection can build on your progress while you sleep to further develop your ability to fulfill your potential.

Satisfied Need for Reaching Potential: Dreaming of Fulfilled Potential

In this extraordinary dream, we realize our potential. We strive to be our best self and achieve this singular goal. However, our accomplishment does not inflate our pride. Instead, we experience a profound sense of contentment. (Dreams of excellent performance, in which we master hard tasks are discussed in Satisfied Need for Mastery: Dreaming of Excellent Performance, page 179. Dreams of excelling at basic tasks by "having the touch" are described in Satisfied Need for Adequacy: Dreaming of Competence, page 183.)

A man in his forties dreamed of achieving fulfillment:

> *It was [an] out of body experience. Woke with feelings of ultimate satisfaction. The dream contained no one except [me] lying on a slab, but not [in a] mortuary. I was looking at myself within the dream. I was pain-free and floating, looking back at myself.*[24]

Self-actualization dreams can also be overtly spiritual. Janet was an arts professional who returned to school to advance her career.

During this period, she found herself at a crossroads between entering a Ph.D. program and accepting an administrative position. After accepting the latter and becoming an executive director, she dreamed:

> *I'm having open heart surgery and I'm lying on my back in an operating room and, as I'm lying there, I'm aware that there's a long and snaking line of surgeons. So, each individual person steps up, places [their] hands on my heart and makes some adjustment, and then they move on. As I look at them more carefully, I realize that these are people that I know. There are family and friends and teachers and people who have been important in my life. And they are both people who are living and people who have passed on. And as each one touches me, I feel I am being healed in some way. Coming out of my heart, out of my chest, is this radiant, rainbow light, and it's just filling the room and it's just filling me with such a feeling of bliss, contentment, and I just know that everything is as it should be at this point in my life; I'm moving in the right direction, and I'm surrounded by people who are helping to heal me. And I thought, wow, what a confirmation that I made the right decision and I'm absolutely on the right path in my life.*[125]

Janet took decisive steps toward self-actualization. In her profound dream, emotional selection tested whether her schemas were prepared to cope with her new reality of being an actualized person. She fully embraced her dream, suggesting that she will likewise embrace self-actualization while awake.

Dreams of reaching our full potential are often accompanied by contentment. When schemas effectively process these remarkable dreams well, we embrace our dream fully. When processed ineffectively, we reach our potential, but doubt our ability to remain a self-actualized person.

The satisfaction experienced during dreams of self-actualization may deepen over the course of recurring dreams as emotional selection adapts our schemas to meet this important need. If you are among the rare people who often dream of self-actualization and struggle to embrace important dream moments, you can review your related perceptions while awake. Perhaps as a child, your caretakers did not encourage you to reach your full potential, believing that such pursuits were selfish, or feel undeserving of self-actualizing. Regardless of your perceptions, reconsidering those that inhibit your ability to self-actualize is a step toward enhancing underlying schemas. As your perceptions improve during waking, emotional selection can further adapt related schemas while you sleep to improve your ability to self-actualize.

Other Self-Actualization Themes

Our need to self-actualize can take many forms. For artists, self-actualization may include mastering their creative process. For professionals, self-actualization may include attaining a deep understanding of their field. However, self-actualization entails more than simply excelling at our job. True self-actualization incorporates important aspects of our life, including our relationships with others and ourselves. Regardless of the form taken by dreams of self-actualization, these dreams signify emotional selection is conditioning our schemas to meet our need to realize our full potential.

Transcendence Dreams

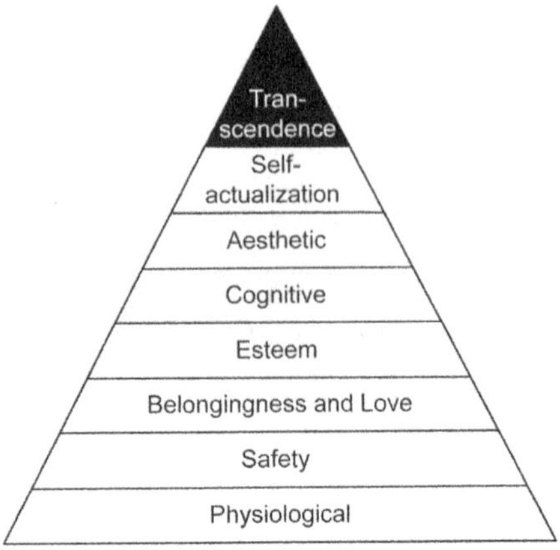

Originally, Maslow positioned self-actualization at the pinnacle of his hierarchy, but later in life, he contemplated the needs of the rare individuals who achieve self-actualization. What drives those who fulfilled all the needs of the self? The answer he found was in fulfilling the needs of others. Maslow proposed that after people reach their potential, rather than becoming passively satiated, they turn their attention toward helping others. According to Maslow, "Transcendence refers to the very highest and most inclusive or holistic levels of human consciousness, behaving and relating, as ends rather than means, to oneself, to significant others, to human beings, to other species, to nature, and to the cosmos."[126]

Maslow found a paradox in transcendence. Although he considered transcendence the capstone of his hierarchy, he noticed that people often transcended without self-actualizing:

> I find not only self-actualizing persons who transcend, but also nonhealthy people, non-self-actualizers who have important transcendent experiences. It seems to me that I have found some degree of transcendence in many people other than self-actualizing ones.[127]

As with all our needs, emotional selection works to improve the ability of our schemas to cope with occasions when our need to transcend is thwarted, and when this highest of needs is satisfied.

Altruism Themes: Failing Others Versus Helping Others

Discussions on altruism may evoke the names Gandhi or Mother Teresa, but the transcended walk among us. Examples include teachers who work selflessly to better the lives of their students, construction workers who build to a level of craftsmanship that only they can see or appreciate, and volunteers who help those in need.

Altruism may be fleeting acts of bravery or chronic movements to better our world for future generations. Regardless, we all benefit from the charitable acts of people who meet their need to transcend.

Thwarted Need for Altruism: Dreaming of Failing Others

In this disappointing dream, we fail to meet the needs of others. We may attempt to instruct a class full of youths or better the lives of our own children, but are hindered. We may attempt to save a group of people in peril, but face an impenetrable obstruction. Regardless of the dream scenario, we wake discouraged and sad.

Anne dreamed of being unable to help desperate animals:

> *I was walking through a dense jungle. It was cool and lush and smelled of exotic flowers ... I slowed as I neared what looked like an opening in the leaves. As I got closer, I found myself at the edge of a cliff-like hill that stood about 30 feet higher than the ground below. What I saw, to my great delight, was elephants bathing in their sacred pool. They were splashing and playing and swimming. It was very wondrous, and I felt blessed to be privy to this event. The dream*

> *shifted and I was on a beach. A procession of elephants garbed in full high-style Indian saddlery—tassels and gold, etc.—came to [the] water's edge. Then their handlers slaughtered them. I woke up weeping. It was an extremely profound and deeply distressing dream … and I felt sorrow and helplessness.*[24]

Sometimes we can help others in our dreams. Other times, we find ourselves powerless. To be altruistic, our schemas must prepare for times when our attempts prove futile.

Dreams of thwarted altruism do not need to be as abstract and overtly spiritual as Anne's. Susan, a former inmate in her fifties, dreamed of being back in prison and trying to help her fellow inmates:

> *I dreamed I went back to jail, and there I saw a few inmates in my dreams. I also saw some who were sent back to the jail, and they were telling me how bad prison was, that they didn't get treatment and wanted me to help them. I was awoken when I was saying I can't help, you have to follow what you are supposed to in the rules. They wanted me to help and I could not help.*[24]

In Susan's realistic dream, inmates turned to her for help. She did her best by advising them to follow the rules to meet their needs, but her advice was ignored. They expected more from Susan, but she was unable to help her fellow inmates.

Dreams of an unmet need to transcend can be disheartening. When schemas effectively process dreams of thwarted altruism, we feel deeply disappointed at our inability to help those in need, but understand that our failure does not reflect on our ability to help others in the future. When processed poorly, we believe our inability to help others reflects on our character, or that someone more capable would have succeeded.

The disappointment that accompanies dreams of failing others may ease through recurring dreams as emotional selection adapts schemas to cope with occasions when our need to transcend is unmet. If you find that you blame yourself for failing to help others during these dreams, you can review your perceptions for parallels in your waking life. Perhaps an influential person in your past instilled insecurities that challenge your ability to help others, or a caregiver did not support your need to transcend. Regardless of your perceptions, reconsidering those that inhibit your ability to help others is a step toward enhancing underlying schemas. As your perceptions improve while awake, emotional selection can further adapt your schemas while you sleep to enhance your ability to transcend.

Satisfied Need for Altruism: Dreaming of Helping Others

In this deeply satisfying dream, we succeed in helping others. They may be coworkers, family, or friends, or they may be complete strangers, animals, or spiritual beings. Rather than expecting gratitude, our reward is knowing that we bettered the lives of others.

Barb, whose dreams we read earlier, was an oldest child, so she was likely accustomed to catering to the needs of her younger siblings. She dreamed of transcendence in the context of nature and people:

> *I'm in a swimming pool. There are seals and an octopus in there. At first, I'm a little afraid of the octopus, but they all come to me and I learn to love them. They are sweet, loving creatures who look at me mutely with soft, big, moist eyes just radiating love and [the] need for love. I stroke them on the head and pet them. They love it and need it. I get out of the pool and one creature follows me. He is tiny and so needy.*

> *His name is Tom. He curls up on me and I pet him and soothe him. He's so small and needy. At one point he's lying on my thumb and gets absorbed into my skin. At first, I am concerned, but then I realize he needs that nourishment from my body. He later comes back out, large and more happy. Now I'm a teacher in a rural, small, one room high school. It is graduation day. My class is seated on the bleachers and I stand before them. I am touched by their beauty, intelligence, loving natures, and creativity. I give them a speech. I say, "I know you are feeling sad at leaving your dear friends of four years. I, too, am feeling sad, for never have I seen such a loving, creative, intelligent, beautiful group of students. If you are the typical graduating class, then this world will be blessed and things will change. You go out, each of you, melded with your friends. A piece of them lives in you now. So, bless you and good luck." I begin crying. Someone shuts the door on them as soon as they are out, but I open the door again and continue my grieving of their departure. Then I stop crying and feel peaceful and loved.*[128]

Barb met the needs of the creatures and students of her dream. She transcended the needs of the self and put the needs of others first, fulfilling her need to transcend.

Not all dreams of transcendence are as abstract or dramatic as Barb's. In his fifties, Steve had a simple yet poignant dream of anonymously helping others:

> *I dreamed that I was a large lottery winner and I had kept it a secret from every one of my friends and family. I just was secretly helping some people with things that mattered to them. And the great enjoyment I got from doing it was beyond words.*[24]

In Steve's dream, rather than spending his good fortune on himself, he gifts it to others anonymously, so receives no gratitude from the beneficiaries. Rather, his reward is knowing that he fulfilled the needs of others. He expresses the feeling elicited from his act as "beyond words." His experience appears to be truly transcendent.

Dreams of altruism are often profound. When our schemas process dreams of transcendence well, we feel a deep sense of contentment from helping others meet their needs. When processed ineffectively, we may feel pride instead of contentment, or feel unworthy of transcendence despite helping others.

The gratifying feelings that well during dreams of transcendence may increase over the course of recurring dreams as emotional selection improves related schemas. If you frequently dream of transcendence and find that you struggle to embrace dream moments, you can review related perceptions while awake. You may realize that your caregivers raised you to focus on your personal needs, rather than the needs of others, or discover that you overlook your many acts of kindness. As you reflect on your perceptions and underlying schemas, you may acknowledge that you already help others meet their needs. For instance, they may have safety needs, and you put yourself in harm's way to protect them, or they may have belongingness needs, and you help them find love. As you reconsider your perceptions and your schemas improve, emotional selection can build on your progress while you sleep to further develop your ability to meet your need to transcend.

Other Transcendence Themes

Dreams of transcendence can take many forms. In the above dreams, Barb gifted her body to creatures and her time and knowledge to her students, while Steve used his newfound wealth to meet the needs of others.

Transcended people help others meet virtually any need, from their most primitive needs to their highest needs. Consequently, dreams of transcendence can include a wide range of themes, from helping people meet needs for food, shelter, and safety, to their needs for belongingness, love, knowledge, creativity, aesthetics, self-actualization, and transcendence.

Regardless of the scenario of your transcendence dream, you can review the emotions, feelings, and perceptions evoked during your dream for parallels in your waking life. As you gain insights into the similarities between your dream and your experiences while awake, you will be influencing underlying schemas. Emotional selection can build on this progress while you sleep by modifying and testing your schemas with dreams, which will further improve your ability to transcend.

Omitted and Miscategorized Dreams

The above sketch is far from exhaustive. A comprehensive description of human need categories and the dreams that modify and test the schemas that help us meet these needs would easily fill a bookshelf. So, you may be disappointed that this chapter did not deconstruct a vivid dream of yours. Fear not. As discussed, you can use emotional selection to deconstruct virtually any dream theme. Although the examples in this chapter are not comprehensive, the techniques described above are.

Consider a dream in which someone ignored your important contributions. In your dream, you worked alongside a coworker to complete an important task, but he or she took all the credit. You awoke feeling taken for granted and sad. Deconstructing your dream, you conclude that you felt unappreciated, yet you find the category "appreciation" missing above. However, you already performed the most important part of dream deconstruction—you identified the perceptions and feelings you experienced during your dream.

Digging a bit deeper into your dream of being unappreciated, you can evaluate how effectively your schemas performed. If you initially felt sad in the dream, but subsequently realized the lack of appreciation you received does not reflect the importance of your contribution, that is a good sign your schemas coped well. If your feelings were overwhelming, they signify your schemas coped poorly with being unappreciated. Reviewing your perceptions while awake can help identify those that are maladaptive. You may find you place too much emphasis on the helper side of your personality, or require recognition for every accomplishment. Regardless of your perceptions, reconsidering those that limit your ability to cope with occasions when you are unappreciated is an important step toward identifying and enhancing related schemas. As your schemas improve during waking, emotional selection can build on your progress while you sleep to further improve your ability to weather situations in which your need for appreciation is unmet.

A challenge with categorizing dreams is that dream themes often span multiple categories. For instance, dreams of poor performance were subcategorized above under esteem dreams, because of the diminished self-esteem that often accompanies these dreams. However, dreams of poor performance often include storylines in which our knowledge is lacking, such as forgetting our lines during a theatrical performance. So, an argument could be made that some dreams of poor performance should be filed under cognitive needs, rather than esteem needs. Consequently, during dream deconstruction, we must be fluid when categorizing the needs that appear in our dreams. If a dream of poor performance feels more like a test of our ability to cope with unmet cognitive needs, instead of esteem needs, deconstruct it as a cognitive dream. If it straddles both esteem needs and cognitive needs, or perhaps another category entirely, then deconstruct the dream as falling under all of these needs. Categorization is not the ultimate goal. Instead, the aim during dream deconstruction is to understand the needs our schemas are adapting to meet through dreams.

Discussion

Despite being the longest chapter in this book, I fear it is overly brief. The topic of human needs is vast, so an adequate conversation on the dreams that emotional selection employs to improve our ability to meet these needs should be equally vast. However, this chapter is not an attempt to categorize every dream theme. Instead, it demonstrates how emotional selection can deconstruct virtually any dream to identify the needs our minds are trying to meet.

The above sketch reviewed how dreams can be deconstructed in the context of Maslow's category of needs. During this exercise, we climbed Maslow's hierarchy, stopping at each rung to review the role dreams play in helping our schemas meet specific needs. However, as mentioned, using Maslow's categories was my personal choice. When deconstructing your dreams, you need not adhere to Maslow's categories, or to any need categories. Instead, the essential goal of dream deconstruction is to understand the relationship between your dreams and your waking needs.

Dream deconstruction is the process of reflecting on your dreams and reconsidering related perceptions. As a dreamer, you are the sole witness to your dreams. You are the only person who experienced your dream, so you are the ideal person to deconstruct it. As you deconstruct the relationship between your dreams and your waking life, your schemas accommodate these insights. Emotional selection can reward your efforts by further enhancing your schemas while you sleep to improve your ability to meet your needs.

This penultimate chapter concludes our discussion on dream deconstruction. In the next, final chapter, I offer a brief opinion on the significance of emotional selection, and an equally brief summary.

NINE

The Takeaway

I have had dreams and I have had nightmares, but I have conquered my nightmares because of my dreams.

—Jonas Salk

THANK YOU for joining me on this exploration of emotional selection. Within the pages of this book, we delved into many topics, including bizarreness in dreams, recurring dreams, systems within the mind, and the role dreams play in modifying and testing our schemas while we sleep.

If you were to take away an idea or two from our discussions, which would they be? For me, this question brings two thoughts to the fore. First, to say that dreams are profoundly powerful is an understatement. By modifying and testing our schemas, dreams literally change our minds. I find this observation truly remarkable. When we wake from sleep each morning, we are transformed through our dreams. Second, we can deconstruct our dreams to identify the perceptions and schemas our minds worked on improving. Using this insight, we can challenge the accuracy of our perceptions. As these perceptions improve while awake, their underlying schemas also improve. Emotional selection can capitalize on our waking improvements to further enhance our schemas while we sleep.

In the introduction, I promised arguments and findings that support emotional selection. I sincerely hope I surpassed your expectations. With your permission, I will spend the balance of this chapter summarizing emotional selection, sans the detail of previous chapters.

Recap

Throughout childhood and adulthood, our schema networks grow in size and complexity, providing a wider and deeper understanding of our world and our place in it. Unfortunately, this expanding knowledge comes with a cost—large schema networks invariably contain maladaptive schemas that conflict with healthy ones, which reduces their efficacy. Thankfully, this is where emotional selection steps in. Emotional selection is a two-phased process that uses dreams to modify and test schemas to improve their fitness for waking life.

The two phases of emotional selection resemble the two phases of natural selection. Darwin's theory describes a variation phase, in which the characteristics of a species vary, and a selection phase, in which nature culls unfit offspring. Emotional selection similarly cycles between a variation phase, in which schemas accommodate new understanding, and a selection phase, in which emotions factor into which accommodations are culled. However, the similarities between emotional selection and natural selection end there. First, natural selection evolves a species on a relatively glacial time scale, cycling once per generation, while emotional selection cycles several times during a single sleep. Second, natural selection employs a life-and-death struggle to cull maladaptations, while emotional selection uses sleep and dreams.

Emotional selection adapts our schemas to our waking lives, but what does it mean to adapt our schemas? Simply put, emotional selection enhances our schemas to meet our needs. We all have needs to be safe and secure, to be loved and to love, to be intelligent and knowledgeable, to be insightful and creative, and to be accomplished and generous, to name a few. The fitter our schemas, the more likely we are to satisfy our needs. So, to improve our schemas while we sleep, our minds process dreams with themes our mental schemas accommodate. These themes may be of us performing well at work, sharing a moment of closeness with a loved one, or enjoying a beautiful sunset. Whatever the dream, our schemas accommodate its theme. But is this new accommodation truly an enhancement, or did it inadvertently diminish our ability to meet our needs? To answer this question, after the accommodation phase, the dream process moves to the emotional selection phase, in which dreams serve as test scenarios that tax prior tentative accommodations. These dream tests can be unpleasant, including themes of being chased, falling, being rejected, and performing poorly. Such unpleasant themes test schema modifications that cope with thwarted needs. Conversely, emotional selection also processes dream tests with pleasant themes, such as flying, finding treasure, making discoveries, and achieving success. These enjoyable themes ensure schemas can embrace occasions when our needs are satisfied.

Bizarre dreams are often our most emotional. This observation is not coincidental. Consider the joy of suddenly discovering we can fly or the terror of being chased by a ravenous Tyrannosaurus rex. While these dream scenarios are bizarre, our emotional responses are not. For instance, we do not experience joy when chased by a Tyrannosaurus rex in our dreams—such a response would truly be bizarre. Rather, the terror experienced during a dream tests whether our schemas can cope with terror while awake. By focusing on the emotional experience of the dream, rather than its bizarre scenario, even our most bizarre dreams become surprisingly rational.

Maladaptive schemas limit our ability to meet our needs, which is why emotional selection targets them during sleep. Unfortunately, such schemas can be deeply entrenched within our mental models, making them difficult to modify. As emotional selection continually works to improve these schemas, it runs similar dream tests, resulting in dream themes that may recur over months, years, and even a lifetime. Regardless of the time frame, we should see a progression toward meeting these needs over the course of recurring dreams, as emotional selection succeeds in adapting our schemas, resulting in more generalized dreams that ensure the schemas can cope with more generalized needs.

A dream theme that recurs without progression signifies emotional selection is stuck. We can help to free the emotional selection process by reviewing recurring dreams for perceptions and needs. For instance, if we repeatedly dream of unmet safety needs, we can review related perceptions while awake. Such reconsiderations may lead to a realization that we are indeed unsafe during our waking lives, and must make adjustments to ensure our safety. Alternatively, we may find that we are as safe as realistically possible and must accept that no need in life is guaranteed. Regardless of our perceptions, as we reconsider our ability to cope with met and unmet needs, our schemas accommodate these insights. As our schemas improve while awake, emotional selection can build on this momentum while we sleep.

During our daily routine, we fret about paying bills or meeting deadlines at work. However, while we sleep, we dream of being reunited with past loves, viewing a breathtaking landscape, or falling from a great height. So, while we may consider the needs that preoccupy our mind while awake as paramount, our dreams reveal the needs our mind is truly concerned with meeting.

Before discovering the theory of emotional selection, I suspected dreams were important due to the large amount of time and energy our mind devotes to dreaming. Despite this preconception, emotional selection truly astounded me. With every sleep, our mind evolves through dreams.

Appendices

Introduction to Appendices

Minds are like parachutes; they work best when open.

— Thomas Dewar

HAS OUR DISCUSSION on dreaming piqued your interest in related topics of the mind? If so, let us continue our journey within the pages of these appendices.

Emotional selection is a relatively new theory. However, it stands on the shoulders of well-established, age-old concepts. As we will discuss within the balance of this book, several aspects of psychology, psychiatry, neurology, and philosophy lend support to emotional selection. These disciplines describe both the pitfalls and paradoxes of the mind, including inaccuracies in our perceptions and how the mind distorts reality to adapt to waking life. During our review of these disciplines, I will answer questions such as, why does the mind intentionally distort perception? What are the distinctions between the brain and the mind? And what exactly is the mind?

The theory of natural selection predated our understanding of DNA and the genome because Darwin did not require insight into the underpinnings of his theory to fully understand its implications. Likewise, the nitty-gritty details of emotional selection are not required for the reader to fully comprehend and apply the theory. However, delving into the underlying mechanisms of any theory can deepen our understanding of both the theory and its far-reaching implications. Consequently, these appendices are provided for those interested in reviewing additional evidence that supports emotional selection, or who harbor an interest in the principles of systems and the mind that paved the way for the emergence of the theory of emotional selection.

These additional pages are certainly supplemental. At this point in your reading, you have a firm grasp of the emotional selection process and how to apply it. However, if you are interested in a deep dive into the foundation of the theory, please consider these appendices as an invitation to continue our exploration into the mind.

Perception Deception

There is no truth. There is only perception.

— Gustave Flaubert

Our perceptions guide our understanding of people, places, objects, and ourselves. However, what we perceive as truth is often remarkably inaccurate. And, as we will discuss in this appendix, that is a good thing.

Types of Perception

Our brain and mind comprise a sensory processing assembly line that continually assigns meaning to our surroundings: that is a person, this is a book, that is a lamp. This assignment of meaning to information from our senses is called *sensory perception*. After identifying objects, schemas characterize them—a process called *cognitive perception*: that person is my friend, this book is boring, that lamp is off, and so forth. The last stop on this perception assembly line is "situation perception," in which meaning is assigned to situations such as, wow, I am being rude because I am reading this book while my friend is visiting.

As the name implies, situation perception processes information to create a "story." Insignificant situations typically do not reach our consciousness, such as entering a room and noticing a lamp on the table. We may see the lamp, but unless its presence holds significance, it may remain in our unconscious.

The perception assembly line feels instantaneous, but it happens in stages and takes time, as keenly demonstrated by a hypothetical visit from your friend.

A friend is in your neighborhood, so she drops by your place for an unannounced visit. After stopping at your front door, she prepares to knock. Inside, you collect your belongings and are about to run some errands. You check your keys, your phone, and your wallet, then turn the knob and swing open the door.

Your sensory perception detects a figure. Even though your visitor is a close friend, you have not processed that information yet. Rather, you perceive yourself as about to be attacked. Certain that your life is in danger, panic takes hold.

Your schemas continue processing the situation. Your cognitive perception identifies your friend's stance—with her fist in the air, about to knock, she appears menacing, further exacerbating your perception. At this point, you blanch or gasp as your adrenaline level rises, your heart races, and your muscles tense.

Eventually, your situation perception kicks in and you perceive your friend and her intentions. As your heartbeat returns to normal, you share a laugh over the gamut of perceptions you displayed.

When discussing perceptions in the context of dreaming and emotional selection, we are typically referring to this last step in the perception pipeline—situation perception. The daily scenarios that schemas actively interpret are often complex and ambiguous, so our situation perception continually evolves to navigate them. For instance, is someone observing us from across a crowded room threatening or friendly? If the latter, is his or her behavior flirtatious or platonic? Schemas drive the perceptions that assign meaning to the complex situations we routinely encounter throughout our day.

Perception Versus Reality

While it feels as though our mind is continually working toward an accurate perception of our world and who we are, this perception is not supported by the evidence. Consider a College Board questionnaire that asked 829,000 high school students to rate themselves in various categories, including their ability "to get along with others." Less than one percent of the students rated themselves as "below average," with 89 percent rating themselves as "above average." Adding further impossibility to the statistics, 60 percent of the students rated themselves in the top 10 percent, and 25 percent in the top one percent—2400 percent higher than statistically possible.

As these students demonstrated, our perceptions often do not reflect reality. Tali Sharot, author of *The Optimism Bias*, clarifies:

> Our brains have evolved to overpredict future happiness and success because, funnily enough, doing so makes health and progress more likely. Understanding how the mind generates and maintains unrealistic optimism and how–even more bafflingly–optimism leads to professional and personal success requires intimate knowledge of how the human brain works. The tendency for positive predictions to create positive outcomes (whether subjective or objective) is rooted in fundamental rules governing the way the mind perceives, interprets, and alters the world it encounters.[129]

Holding inflated notions of our abilities increases our confidence, which, in turn, improves our ability to accomplish tasks. So, while believing a distorted version of reality seems counterproductive, such beliefs can be self-fulfilling.

Emotional selection modifies and tests the schemas that characterize these blissful descriptions of ourselves. If meeting your needs requires an inflated sense of self, then emotional selection tirelessly works to distort your perceptions.

Researchers refer to this tendency to distort our perceptions in a certain direction as *cognitive bias*. Psychologists David Dunning and Justin Kruger famously showed how cognitive bias affects cognition. The inspiration for their research came from an unlikely source: McArthur Wheeler, a bungling bank robber who robbed two banks without wearing a mask. How did Wheeler's exploits influence cognitive researchers? Well, immediately after his crime spree, local news distributed footage from surveillance cameras. The police arrived at his doorstep within hours.

"But I wore the juice," Wheeler responded, incredulous that the police had identified him. It turns out that Wheeler put lemon juice on his face before the bank robberies. He reasoned that, because lemon juice could be used as invisible ink, it would also make a human face disappear. The only downside of his scheme, in his mind, was that the juice stung his eyes so badly that he could hardly see. In his defense, he tested his hypothesis before his crimes by wearing the juice and taking a selfie. The absence of his face in the photo reinforced his conviction, though perhaps he simply missed seeing his face with the camera because of his lemony vision.

Where the general public saw a bumbling criminal, Dunning and Kruger saw cognitive bias. Specifically, they hypothesized that people with below-average knowledge and skills erroneously appraised their abilities as superior to their peers. They tested their idea by inviting novices to try a variety of challenging tasks ranging from reading comprehension to tennis. The participants then rated their competence and that of their peers. Repeatedly, they found newbies were oblivious to their limitations and failed to recognize the competence of others. Only after receiving training did participants acknowledge their lack of skill.

The inflated self-image that accompanies incompetence gained the moniker the *Dunning-Kruger effect*. The researchers also observed the converse—highly skilled people often underestimate their abilities while overestimating those of their peers.

Situation perception assigns meaning to limited, and often ambiguous, information. During conversations, we interpret body language and other nuances to gain further insight. So, we weigh the significance of influential people saying, "I am interested in working with you," with subtleties in their delivery. If they utter the statement while checking their watch, their sincerity is called into question. When stated emphatically while making eye contact, we draw the opposite conclusion. Regardless, our schemas are responsible for navigating these complex situations and arriving at a conclusion.

Schemas also affect our interpretation of past events. Researchers asked a group of college students to teach mathematics to a class of elementary school children. When the college students were later told the children performed well, the students were quick to attribute the success to their teaching skills.[130] When told the children performed poorly, the college students attributed the disappointing results to the limited abilities of the children or inferior classroom conditions. The researchers concluded that people tend to attribute success to themselves while deflecting failures to external factors.

Our self-image plays an important role in how we interpret our surroundings. One study of college students found a strong tendency to distort questions, either positively or negatively, based on the student's self-esteem.[131] At a critical point during an interview regarding mundane college experiences, researchers asked the students personal questions designed to elicit a strong negative or positive self-image. Those with high self-esteem readily answered questions that elicited a positive self-image, but often reworded the questions designed to give negative responses into questions that elicited more positive responses. Students with low self-esteem did the opposite, freely transforming the questions designed to give positive responses into questions that resulted in negative responses.

Our surroundings influence the schemas that characterize our self-image. A study asked a group of participants questions designed to elicit introverted responses, such as, "What things do you dislike about loud parties?"[132] A second group was asked questions designed to elicit extroverted responses, such as, "What would you do if you wanted to liven things up at a party?" Toward the end of the rigged questionnaire, the researchers asked participants whether they were socially introverted or extroverted. Those who answered questions designed to elicit introverted responses were more likely to rate their sociability as introverted. The effect continued after the questionnaire was completed. The participants were asked to take a chair into the waiting room, where a second person, an actor who was part of the study, pretended to also wait. People who answered questions designed to elicit extroverted responses were more likely to place their chair closer to the actor and start a conversation than those who answered the introverted questions.

Perceptions of inflated capabilities have obvious benefits, but can a low self-opinion also have benefits? Psychologists suggest that low self-perceptions can reduce anxiety and support relationships. Children with abusive parents could develop an accurate perception that their parents are inadequate, but the resulting fears and anxieties of having bad parents are frightful for a child. Instead, such children typically develop the schema that their parents are good and it is they who are flawed and undeserving of love.[133] This inaccurate perception preserves the parent-child relationship, resulting in reduced anxiety compared to developing perceptions that accurately reflect the child's situation.[134]

When generating a perception of reality, our schemas strike delicate balances between accuracy and exaggeration. While we sleep, emotional selection experiments with these delicate balances to optimize our schemas for our waking lives.

Discussion

Dreams often rework our schemas to generate distorted perceptions of reality, and that can be a good thing. By distorting our perceptions of ourselves and others, we increase our motivation and our ability to cope with life's challenges.

Emotional selection describes dreams as modifying and testing the schemas of the mind. While this description of dreaming is simple, schema networks are large and complex, with schemas that form early in childhood often becoming entrenched, making schema modification and testing difficult. In the next appendix, we will explore the challenges that emotional selection faces while modifying and testing our schemas through dreams.

Changing Your Mind

I can never decide whether my dreams are the result of my thoughts, or my thoughts the result of my dreams.

— D. H. Lawrence

WHEN YOU OPEN your kitchen freezer, moisture, carried by air, condenses and freezes as frost on the surfaces within. To prevent frost from accumulating, "frost-free" freezers periodically raise their internal temperatures to slightly above freezing to melt newly formed frost. The melt drains through a tube to a pan underneath the refrigerator, where it evaporates.

According to a theory by Giulio Tononi, our brain uses a process that is a bit like frost-free refrigeration. During the day, the brain forms new neural connections. Some connections support new memories or newly learned skills. Others are incidental, created by the day's activities, and must be purged, like frost in a freezer.

How are these unwanted connections "melted?" During non-REM slow-wave sleep, the brain repeatedly fires its neurons for half a second and then rests them for half a second. For complex bioelectrical reasons, this weakens the connections between neurons, eventually purging the brain of its abandoned neural connections. This indiscriminate weakening of neural connections "primes" the brain for improved learning by increasing its plasticity. In other words, after non-REM sleep, your brain is more malleable.[135]

The increased plasticity during non-REM sleep creates the ideal conditions for emotional selection. As discussed, emotional selection maps to the sleep cycle—accommodations occur during non-REM sleep and dream tests occur during REM sleep. So, priming the brain for learning during non-REM sleep provides the perfect conditions for developing new accommodations.

In this appendix, we delve deeper into the relationship between emotional selection, your brain, and your mind, and explore the challenges schemas face when adapting to waking life.

Dream Improv

In 1970, students of William Dement, an early pioneer in sleep and dream research, conducted a simple experiment. Study participants slept with electrodes that monitored their sleep stages. Ten seconds into REM sleep, the researching students played familiar prerecorded sounds, including a rooster crowing, a steam locomotive (complete with whistle), a bugle playing reveille, a dog barking, traffic noise, and a speech by Martin Luther King. They then stopped the tape, woke up the participants, and asked what they were dreaming.[136]

The sounds greatly influenced the content of the dreams, with the locomotive sound being the most effective, influencing 82 percent of the dreams collected:

> *I dreamed I was riding in a train. I was driving the engine, and the train was in Branner dormitory, and right close to the engine there was this pit. It was about two or three stories long, and it was still open, and the train kind of chugged down into it, and it was real[ly] scary. I was dreaming the whole time. When I was going into the pit, it was amazing, because there were some people at the top of the pit watching me go down.*

The introduction of external stimuli into participants' dreams poses a well-known challenge for researchers. The lab environment, equipment, and electrodes affixed to participants' heads appear in roughly one third of collected dreams:

> *I dreamed I was lying here and something went wrong, so that [at] any second I was going to be electrocuted. I wanted to tear the wires off, but suddenly realized that my hands were tied. I was very relieved when you woke me up.*[137]

External stimuli have less effect on the second night in the laboratory. So, researchers often collect dreams for two or more evenings, discarding those from the first night's sleep.

Our ability to incorporate train whistles and other stimuli into the storyline of a dream resembles the improvisation that actors often use in live theater. During a theatrical performance, external stimuli in the form of broken props and actors forgetting lines are common. The director could address such interruptions by stopping the performance, apologizing to the audience for the disruption, addressing the issue, and then restarting the performance. However, this approach would take the audience out of the scene that the actors took great pains to create. Rather than stopping, performers improvise. As the saying goes, "The show must go on."

Consider one such live performance of the comedy, *The Pig Farm*. During a scene, an audience member's mobile phone's melodious ringtone played loudly and repeatedly. Rather than allowing the interruption to remind the audience of their surroundings, the actors enveloped it. "My God, that's beautiful," an actor said, still in character. "What is that?" His accompanist responded in kind, "On a clear day, chimes carry over the mountains." The actors improvised for another two minutes, culminating with, "Anyway…," then returned to the script. Much of the audience thought the improvisation was a planned part of the performance.[138]

A dream is a virtual stage, complete with scenes and actors. Our mind is unaware of how we will react to unfolding dream themes. Dreaming of being chased, we may flee or turn and face our pursuer. Our mind does not know. Instead, our mind is the improvisational director of the dream, so it monitors developments and improvises.

While dreams often incorporate external stimuli, do examples exist of the mind using improv to incorporate stimuli generated from within the mind? That appears to be the case in the findings of Erin Wamsley and Robert Stickgold.[139] To investigate the interplay between dream content and memory consolidation, the researchers designed a clever experiment in which participants played a 3-D computer maze game. The researchers then gave the participants a five-hour break, in which one group of game players napped and the other group stayed awake and sat quietly in a room.

During the break, the researchers checked in with the awake group and asked what they were thinking about.[140] They also awoke the napping group during non-REM sleep and asked each of them, "What was going through your mind?" Slightly more than half of the participants' dreams were relevant to the task:

> *I was thinking about the maze and kinda having people as checkpoints, I guess, and then that led me to think about when I went on this trip a few years ago and we went to see these bat caves, and they're kind of like, maze-like.*
>
> *Looking for something [in a maze].*
>
> *Just hearing the music [from the task].*

After the five-hour break, the participants played the game again. Which group's performance improved the most after the break? Naturally, it was the napping group. Overall, those who slept outperformed those who did not. Furthermore, the nappers whose dreams included aspects of the maze game improved tenfold compared to the other nappers.

The dreams of the napping group were often only tangentially relevant—there were no checkpoints or people in the maze game, nor were there bat caves. However, the maze experiment clearly influenced their dream content. Perhaps their minds used improv to incorporate the memory consolidation from the game into their non-REM dream, much in the way that Dement's participants incorporated a train whistle.

Non-REM sleep supports emotional selection, memory consolidation, and other processes. The brain is a multitasking powerhouse, capable of supporting these processes independently and concurrently. These processes likely bump into each other occasionally, resulting in memory consolidation influencing dreams.

Mental Model

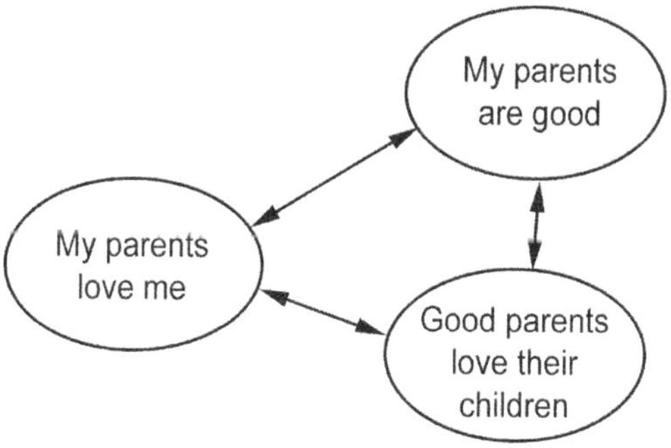

Schemas support a mental model of ourselves and our world. To gain insight into the interaction between schemas, consider two schemas in the above figure that a child may develop early in life: "My parents are good" and "My parents love me." By finding commonality in existing schemas, new generalized schemas form, such as "Good parents love their children." The arrows of the figure below show the interdependencies between schemas, which define their cause-and-effect relationship. In this example, we could say, "Good parents love their children because my parents love me, and my parents are good."

Once established, these generalized schemas support specific ones. For instance, we could say, "My parents love me because good parents love their children, and my parents are good." We could also say, "My parents are good because my parents love me, and good parents love their children." As these interdependencies form, the mental model of our world grows.

Safety is a basic need. So, possessing a sense of safety is critical for children. In our example, the schemas "my parents are good" and "my parents love me" contribute to the new generalized schema, "I am safe." Since the parents of this example love their child, the generalized schema, "I am lovable," also forms, as shown below. As we shall see shortly, such schemas become critical as the child matures and develops meaningful relationships outside of the family unit.

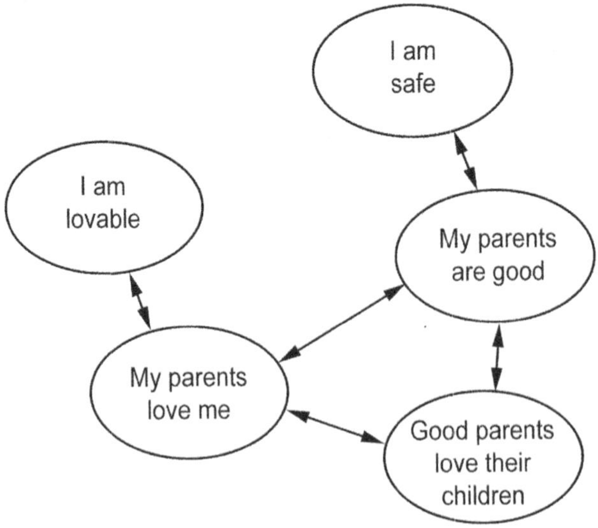

Children raised by nurturing caregivers typically develop healthy schemas that reflect their supportive environments, but what about children raised in mentally unhealthy environments? A child raised by abusive or neglectful parents could develop schemas that closely reflect reality, such as "I am lovable," while also maintaining realistic schemas about their parents' limitations, such as "my parents are inadequate" and "my parents are incapable of love." While accurate, such realistic schemas make children feel unsafe, which can lead to depression and chronic anxiety, resulting in reduced efficacy and motivation. So, children with inadequate caregivers often develop adaptive, albeit less accurate, schemas to cope.

Because the realistic schema "I am not safe" can be debilitating, unnurtured children often develop distorted schemas, such as "I am unworthy of love" and "Good parents love only worthy children," to support their schemas "I am safe" and "My parents are good." So, as shown in the figure below, because children need to feel safe, schema networks may characterize low self-esteem to help a child fit into an unsupportive environment.

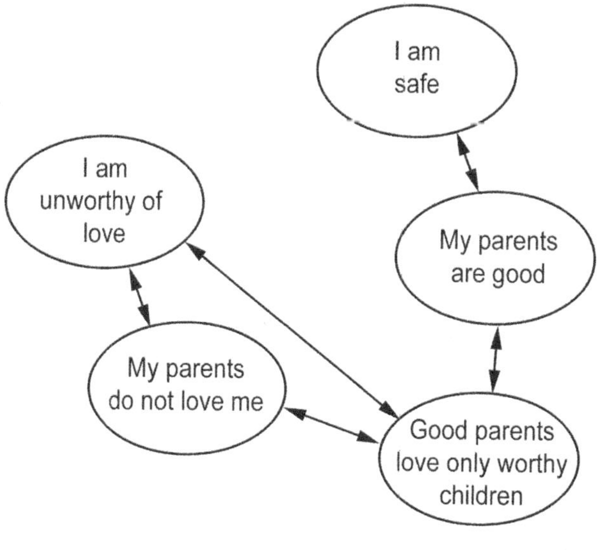

As the child of inadequate parents matures and forms new important relationships, he or she may feel conflicted when shown genuine love. For instance, marrying a loving spouse could create the new schema, "my spouse loves me," which conflicts with the existing one, "I am unworthy of love." As shown below, such conflicts challenge the accuracy of the mental model, triggering anxiety.

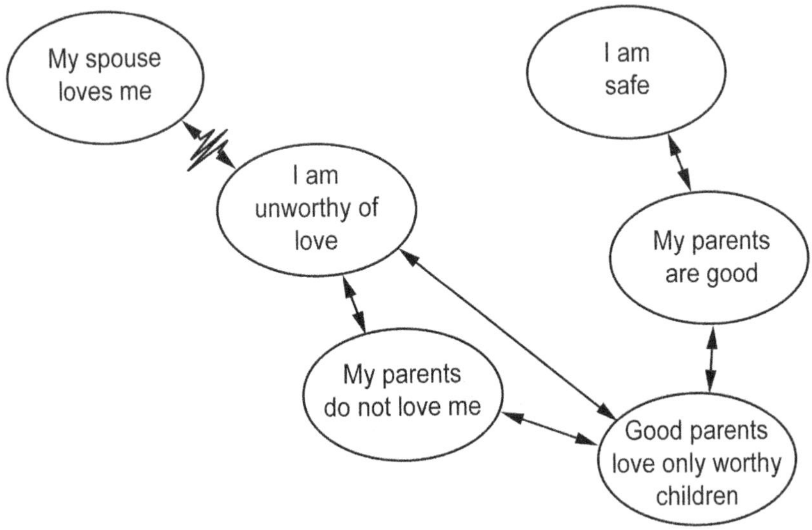

The schema network in our example could adapt to a loving spouse by altering existing schemas to reflect reality. As an adult, a self-image of being lovable is more adaptive than maintaining a schema of being unlovable. Unfortunately, schemas developed during childhood can be deeply embedded in the schema network, making them difficult to modify.[141]

Rather than formulating realistic schemas to accommodate a loving spouse, a new generalized schema often develops: "Those who love me do not know me." This new generalized schema resolves the conflict by allowing the schema "my spouse loves me" to coexist with the schema "I am unworthy of love."[142]

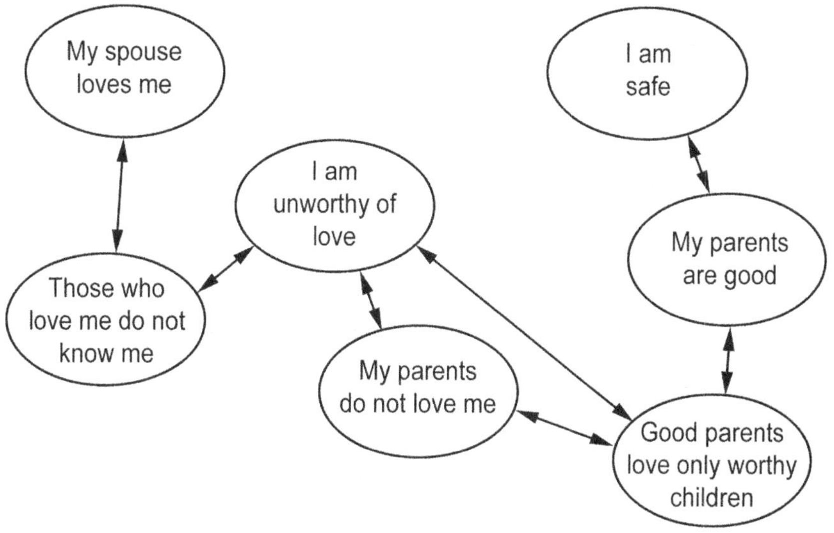

Our schemas often do not reflect reality. This can be beneficial when coping with adversity, but maladaptive when inaccurate schemas prevent us from forming loving, supportive relationships. Fortunately, schemas such as "those who love me do not know me" often activate alongside adaptive ones, such as "my spouse loves me." As we discussed in Part Two, listening to our adaptive schemas helps to heal maladaptive ones.

Mental Town Hall Meetings

Emotional selection routinely chooses which schemas to modify and test. Given the countless schemas in a mental model, how does emotional selection make this choice? A simple answer is a mental process resembling a town hall meeting. As an analogy, imagine a hamlet assembling in a large room with folding chairs and wooden floors. Attendees include farmers struggling to transport grain to their silos due to poor road conditions, teachers needing new books for their students, and elderly citizens requesting a new roof for their senior center. The community gathers to form a consensus on where to allocate limited resources. Occasionally, someone speaks as the representative of a group. No single group fully comprehends the needs of the others, which spurs sporadic murmuring as people talk among themselves about the issues at hand.

Schema networks likely operate much like a town hall meeting when determining which schemas to modify during sleep. Collectively, schemas characterize everything we know, but individually, they contain a limited understanding of the outside world. This discrete understanding requires schemas to signal during sleep to "request" adaptation. To prioritize which schemas receive accommodations through dreams, the urgency of these requests is compared.[143]

Researchers have found high-frequency bursts of brain waves occur during non-REM sleep. Called *sleep spindles*, the sleep sciences have not yet attributed this brain activity to a neurological process. Schemas communicating among themselves to determine which schemas emotional selection should target with dreams, in the fashion of a town hall meeting, would explain the presence of sleep spindles during non-REM sleep.[144]

Basic Emotions and Emotion Schemas

In 1872, thirteen years after *The Origin of Species*, Darwin's book, *The Expression of the Emotions in Man and Animals*, explored the relationship between emotions and behavior: "… young and the old of widely different races, both with man and animals, express the same state of mind by the same movements." Darwin described how emotions help us navigate our life, including protecting ourselves and loved ones:

> No emotion is stronger than maternal love; but a mother may feel the deepest love for her helpless infant, and yet

> not show it by any outward sign; or only by slight caressing movements, with a gentle smile and tender eyes. But let [anyone] intentionally injure her infant, and see what a change! how she starts up with threatening aspect, how her eyes sparkle and her face reddens, how her bosom heaves, nostrils dilate, and heart beats; for anger, and not maternal love, has habitually led to action.

Darwin observed that emotions spur action, help us communicate with others, and are critical for the survival and proliferation of species. We now know that emotions form in the amygdala, an old part of the brain. In other words, the primitive animals from which we evolved also relied on their emotions to navigate their physical and social environments.

Darwin categorized emotions and studied them scientifically. His keen observations influenced Paul Ekman, a psychologist who extensively researched emotion in diverse cultures around the world.[145] Ekman considered emotions to be distinct, physiological, and measurable. In one of his studies, he visited the South Fore culture in Papua New Guinea. The Fore people's culture had not advanced beyond the stone age; they continued to use stone tools and weapons, and they had no written language. Therefore, Ekman was unable to use the technique of showing participants photos of people and asking them to select the matching emotional expression from a predefined list of words. His creative solution was to tell each participant a story with an emotional storyline through a translator. The participants then selected from a group of photos of people expressing various emotions.

Without a connection to the outside world, the Fore people had no access to media, including magazines, film, or television, so could not have learned emotions from Western culture. Despite their isolation, they chose the same photographs as literate cultures.

Ekman concluded that six "primary emotions" consistently radiate as facial expressions in all cultures around the world—anger, disgust, fear, happiness, sadness, and surprise. He also described these basic emotions as blending to form other, more complex emotions. For instance, anger and disgust merge into contempt. Later, Ekman added basic emotions not encoded in facial expressions—amusement, contempt, contentment, embarrassment, excitement, guilt, pride in achievement, relief, satisfaction, sensory pleasure, and shame. Because these emotions are not genetic, they may vary across cultures.

Despite agreement among the sciences on the importance of emotions, emotions are surprisingly hard to define, even for researchers. Psychologist Carroll Izard related this difficulty in his influential article, "Basic Emotions, Natural Kinds, Emotion Schemas, and a New Paradigm." He asked 33 experts across several disciplines and specialties to define emotion:

> Not unexpectedly though, several noted that (a) it was not really possible to define emotion, (b) definition depends on the particular emotion, and (c) definitions would vary widely. As expected, there was indeed no consensus.[146]

Izard differentiates between basic emotions and emotion schemas. He defines basic emotions as natural kinds, meaning they are encoded in our DNA. They include joy/happiness, sadness, anger, disgust, and fear. Izard describes how babies communicate these basic emotions via face, voice, and body language. These emotions evolved to trigger behaviors in caregivers. For instance, a baby's cry of hunger or fear triggers caregivers to respond with food or protection.

Izard likens emotion schemas to the culinary palates we develop as adults. Our palates are simple in infancy—sweet, salty, sour, and bitter. These flavor distinctions protect infants by rejecting certain foods that may be dangerous. As we mature, our cognition comes into play as we learn to enjoy sour tastes that we rejected as children. Likewise, the basic emotions we possess at birth help us navigate infancy, but as our cognition develops, our emotional responses become more nuanced and complex. According to Izard, "The capacities that enable individuals to frame emotions neurobiologically and develop emotion schemas may be built into the neural systems that underlie both emotion and perception-cognitive systems." For instance, as adults, we feel elevated levels of joy or anxiety when meeting influential people, but a toddler has no concept of fame, so their response to a celebrity is the same as meeting a common stranger.

To test whether the brain processes basic emotions and emotion schemas differently, researchers trained rats to anticipate a brief, yet painful, electric shock. Before reintroducing the shock, the researchers mapped the electrical activity in the rats' brains. Following this, the scent of a fox, a natural predator of the rats, was introduced into the enclosures. Again, the researchers mapped the rats' neural activity. As expected, while both experiences elicited fear, different regions of the brain were triggered for each experience, demonstrating neural differences between learned fears (emotion schemas) and instinctual fears (basic emotions).[147]

As discussed, emotion schemas color our experiences, leading to perceptions that can deviate from reality. Sometimes the perception is rosier. Other times, it is more morose. Izard elaborates:

> In anger-prone children (characterized by persistent or trait like anger schemas), activation of an anger schema may cause the child to perceive anger-evoking cues that others see as neutral. Such anger-perception bias tends to make children more likely to engage in unprovoked aggression. Thus, emotion schemas may become complex emotion-cognition-action systems.[148]

Izard characterizes emotion schemas as elaborate and complex. For instance, feelings of morality and self-consciousness exist in rudimentary forms during childhood but morph and color during adolescence and adulthood. So, a toddler may experience raw fear when encountering a stranger and scamper to the nearest caretaker. Such encounters during adolescence and adulthood may trigger more complex emotion schemas, such as self-consciousness or ego inflation.

Emotions are such an integral part of our mental models that Izard describes the term "emotion" as encompassing both basic emotions and emotion schemas:

> I propose that all mental processes are influenced by the ever-present affect or emotion. A growing body of research shows that perception, cognition, decision making, judgment, and action are influenced by emotion.[149]

So, while researchers struggle to define emotion and disagree on its nuances, they unite on one topic: emotions play an important role in our evolution and in our lives.

Feelings Versus Emotions

People often use the terms "feeling" and "emotion" interchangeably, but there are distinctions. Emotions are physiological responses. Feelings are our interpretations of emotions. So, finding ourselves friendless elicits sadness, leading us to feel lonely, while winning the lottery brings about joy, causing us to feel lucky.

The brain processes feelings differently from emotions. As mentioned, emotions incorporate the amygdala—an "old" part of the brain. The neocortical region of the brain, which is larger and newer than the amygdala, then combines these emotions with beliefs, experiences, and memories to create feelings.[150]

Emotions help us navigate complex social interactions. While conversing with someone, we may find ourselves surprisingly emotional. *Why am I so annoyed with this person? Why do I have so little patience right now?* Afterward, we may reflect and realize that we felt exploited, but in the moment, we were only aware of the emotions. Such experiences demonstrate the differences and importance of both emotions and feelings.

Feelings can consolidate several emotions into a singular understanding. We may feel taken for granted by family members or coworkers, but there is no single emotion to describe this feeling. Rather, this feeling of being exploited is a collection of several emotions, such as anger, sadness, and shame. In short, our feelings help us understand our emotions.

Discussion

In this appendix, we discussed the wonderful flexibility and complexity of the mind, and the challenges it faces as our schema network evolves from childhood through adulthood. We also explored emotions in more detail, including the role emotions played during the evolution of our ancestors, and how emotions help us navigate our daily lives.

In the next appendix, we gain a deeper understanding of emotional selection by considering the entity that produces it—the mind. Specifically, what is the mind? And how does it differ from the brain?

Mindful Abstraction

If I look confused it is because I am thinking.

— Sam Goldwyn

When asked by a toddler to define an airplane, you may answer, "It's sort of like a car, but with wings, so it can fly." Your definition incorporates a physical and a functional description. However, when asked to define the mind, its intangibility poses a challenge. What is the mind? You may be tempted to answer, "A bunch of neurons that work together to think," but how does that definition differ from the brain's?

This challenge of defining the mind exists within and across psychology, psychiatry, and philosophy. "Dualists" deem the mind too complex to merely be the product of neurons, hypothesizing that undetected energy aids neurons in constructing the mind. Some theorists adopt an opposing stance, holding that the brain and mind should be considered a single entity.[151] The lack of consensus is vividly demonstrated in Charles Hampden-Turner's book, *Maps of the Mind*, which chronicles an astounding sixty varying definitions of the mind.[152]

As mentioned, theories that depict dreams as byproducts of unrelated neural functions during sleep, such as the sorting of short-term memory into long-term memory, are referred to as brain-based. Opposing theories describe dreams as intentionally created by the mind to improve itself. Because emotional selection sits squarely in this latter mind-based camp, a simple definition of the mind is helpful for our discussion on dreaming. In this appendix, I propose that we regard the mind as simply an abstraction of the brain. As we will discuss, this consideration can lead to powerful insights into the relationship between the brain and the mind.

Brain ≠ Computer

The brain is often compared to a computer. It is not. Computers and the human brain are fundamentally different. However, the comparison does lend some interesting insights, as parallels exist between our brains and their digital counterparts.

In 1936, British mathematician Alan Turing published his seminal paper describing a new form of calculation machine.[153] His theoretical device required a finite library of symbols, a paper tape, a mechanism for reading and writing symbols to the tape, a set of processing rules, and the ability to store its internal configuration. Gaining the moniker "Turing Machine," this description of the world's first modern computer would write a single symbol at a time, advance the tape in either direction, and update its internal state based on the current calculation.

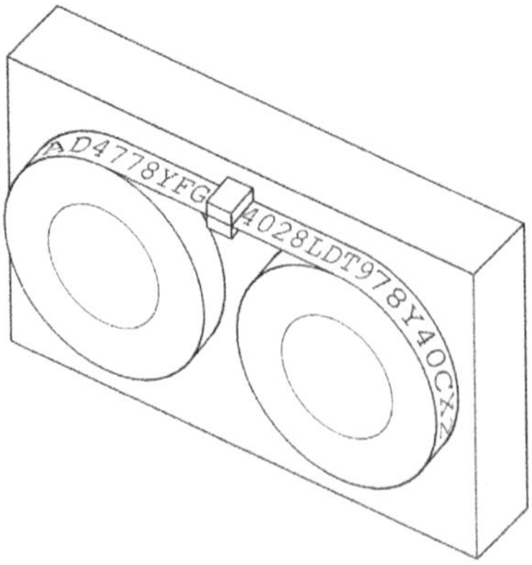

Theoretical Turing Machine

A decade after Turing's paper, the world's first fully programmable computer ushered in the digital age. Named the ENIAC (Electronic Numerical Integrator And Computer), it weighed a staggering 27 tons and was the size of two tractor trailers. Despite being able to hold only 20 numbers in memory, it captured the imagination of researchers. The following year, the field of artificial intelligence (AI) was born. Replete with promises of future machines that could see, hear, speak, and even learn, devices previously limited to the realm of science fiction suddenly seemed plausible.[154]

The first computers performed complex calculations in seconds that took humans days or months. Researchers understood these early machines were merely powerful calculators that could not think. However, they performed so admirably it seemed obvious that further refinement could produce computers capable of complex thought that could rival or even surpass humans.

Psychologists, psychiatrists, and philosophers trying to understand the processes of the mind hypothesized that the mind was simply a symbol-processing Turing Machine. The brain's memory served as the "tape," its neurons as the "circuitry," and the mind as the "rules" or "software." The thinking was that as computer scientists continued developing algorithms for AI computers, we would recognize these same processes in the mind. While neurons bore no resemblance to the hardware circuitry of the day, this observation seemed inconsequential. As the field progressed, theorists believed that such inconsistencies would be clarified.

Psychoanalysts particularly embraced the concept of the mind as a symbol-processing computer, seeing a direct correlation between Freud's description of the mind's use of symbols and the symbols processed by computers. The mind-as-computer analogy also suggested an explanation for our propensity to communicate with symbols, including letters and numbers. We often think of ourselves as rational beings, which may be why computer scientists thought they were on the verge of decoding the algorithms that drive thought.

AI's first steps were groundbreaking. In the summer of 1951, Harvard unveiled a computer capable of "learning" a maze. The more attempts the computer made, the "smarter" it got.[155] Six years later, IBM demonstrated a computer chess game capable of "thinking" four moves in advance. Flush with success, AI researchers made bold predictions. In 1963, the Stanford AI Project was founded with the "goal of building a fully intelligent machine in a decade." Two years later, future Nobel laureate Herbert Simon predicted that "Machines will be capable, within twenty years, of doing any work that a man can do."[156] Such declarations translated into the eager anticipation of domestic robots by the end of the twentieth century. These tireless servants were to do our chores, welcome us home, and even commiserate about our day. However, after missing deadline after deadline, researchers eventually appreciated the towering challenges of building truly artificially intelligent computers.

One does not need to build an AI device to gain a unique appreciation for the complexity of the mind. Neurologist Vilayanur Ramachandran studies patients with missing arms and legs. Some patients lost a limb during an accident, others because of a birth defect. He specializes in those who sense touch in their "phantom limbs." His patients with a missing hand often feel pain when hot water is poured where they perceive their hand to be, or point with a missing digit when giving directions. A common complaint relayed by his patients is that they perceive their missing hand as being perpetually clenched in a fist so tight that their phantom fingernails dig into their phantom palm, causing severe pain.

To treat patients with virtually clenched fists, Ramachandran devised a simple box that uses mirrors to reflect the present, healthy limb where its missing counterpart once was. He instructs patients to insert their hands and open them. As patients open their healthy hand, the mirrors depict their missing hand as also opening. Patients "seeing" both hands opening report their phantom hand as also opening, relieving the sensation of their nails driving into their palm. His remarkable experiments led him to an insightful conclusion:

> These experiments are important because they flatly contradict the theory that the brain consists of a number of autonomous modules acting as a bucket brigade. Popularized by artificial intelligence researchers, the idea that the brain behaves like a computer, with each module performing a highly specialized job and sending its output to the next model, is widely believed. ...
>
> But my experiments with these patients have taught me that this is not how the brain works. Its connections are extraordinarily labile and dynamic. Perceptions emerge as a result of reverberations of signals between different levels of the sensory hierarchy, indeed, even across different senses. The fact that visual input can eliminate the spasm of a nonexistent arm and then erase the associated memory of pain, vividly illustrates how extensive and profound these interactions can be.

Today, nearly a century after the invention of the computer, the field of AI is finally making good on some of its early promises. In 2022, an AI prevailed over human artists to win first place in the digital art category at the Colorado State Fair. College students and professionals routinely turn to AI to generate text for assignments, articles, and even books. While AI has recently made astounding inroads, it can often be surprisingly incompetent. For instance, GPT-4 is a state-of-the-art AI system capable of problem solving and discussing complex mathematics, yet is terrible at tic-tac-toe and often makes arithmetic errors. As one researcher put it, "The thing is both smarter and dumber than any person you have ever met."[157]

Lightning Calculators

Kim Peek, the real-life inspiration for the movie *Rain Man*, was a lightning calculator—a person capable of perceiving large numbers and performing complex calculations in his or her head. Peek had an exceptional memory, having memorized over 9,000 books while reading two pages simultaneously, one with his left eye and the other with his right, typically in ten seconds per page spread. An entire book took him about an hour. Like most lightning calculators, Peek could do complex arithmetic at a young age. He related a day his brother asked him to do a multiplication problem:

> He asked me to multiply something in my head, like 'What is 82 x 82 x 82 x 82?' I just looked at the floor and closed my eyes. My back went very straight, and I made my hands into fists. But after five or 10 seconds, the answer just flowed out of my mouth. He asked me several others, and I got every one right.

The nineteenth century witnessed a heightened interest in lightning calculators. Born in 1867, Jacques Inaudi was a poor shepherd boy who passed the time literally counting sheep. By the age of seven, he could perform five- and six-figured multiplications in his head, without ever being taught to multiply. Like most lightning calculators, he invented techniques for performing calculations. When multiplying large figures, he broke the calculation down into several simpler multiplications and then added them up.

Inaudi could carry on a conversation while performing difficult calculations that took several minutes, hearing a voice with the answer when his mind finished the calculation. He gave public demonstrations in which he requested large numbers from the audience and then played the flute while performing the calculation. At the end of the musical piece, he relayed the answer.

The ability to perform complex mathematical calculations has such obvious benefits, why did evolution not provide this gift to all humans? The answer appears to be in another characteristic common to most lightning calculators—as a group, they have low levels of overall intelligence. Kim Peek was born with the rare birth defect agenesis of the corpus callosum (ACC) and had an IQ of 87, and Jacques Inaudi's gift was limited to manipulating numbers.

While the majority of lightning calculators possess below-average intelligence, there are exceptions. Carl Friedrich Gauss, the eminent mathematician, was three years old when he famously corrected a mathematical error. His father typically paid his workers at the end of every week, tallying their hours while factoring in a bonus. One particular week, he calculated incorrectly, compelling the boy to exclaim, "Father! Father! You've done it wrong. This is the amount." To everyone's amazement, the young Gauss was correct.

The French physicist André-Marie Ampère also demonstrated powerful mathematical skills at a young age, solving complex equations at four years old, before he knew either letters or numbers. As an adult, he made important contributions to science, including the discovery of the relationship between magnetism and electricity—the electrical unit *ampere* (*amps*) is named for him.

Interestingly, lightning calculators with high intelligence typically lose their ability to perceive complex calculations in their head. As Ampère's knowledge of classical mathematics and science expanded, his faculties for lightning calculation declined. The same decline happened to Richard Whateley, a renowned intellectual and philosopher of the nineteenth century:

> My ability to calculate displayed one peculiarity. It showed itself at the age of four and lasted three years. I could do the most complicated additions in my head, and a good deal more rapidly than others who did them on paper, and no one ever found any mistakes in them. But by the time I began going to school, my ability to calculate had vanished, and since then I have always been very weak in mathematics.

The fact that most lightning calculators have below-average levels of conventional intelligence, or lose their unique abilities as their other intelligences develop, suggests there are tradeoffs between their unique capabilities and more common ones. Ramachandran hypothesizes that these limits on intelligence are due to "some specialized brain regions may have become enlarged at the expense of others."[158]

We often regard lightning calculators as gifted. However, the average person has a far more powerful gift—the ability to perceive physical and social situations. While this gift is often overlooked, it is no less remarkable. If a parallel world inhabited by lightning calculators existed, an average person in our world would seem gifted in theirs. He or she would be lauded as having an amazing ability to perceive the intentions of others and derive social conclusions, appearing to be a mind reader. Because we live in a world full of gifted "perceptionists," we often overlook our amazing capabilities.

The minds of lightning calculators lend further insights into our earlier conversation on the differences between computers and the brain. A problem for computer scientists trying to develop a digital mind is that neurons operate very differently from most computers. Neurons signal their neighbors at a rate of approximately 100 pulses per second. This sounds impressive until you compare it to modern-day computers with connectivity rates commonly more than a million times faster. Does this imply that a computer is a million times faster than a human brain? Yes and no. Neurons are slow, but compensate by operating in *parallel*, with millions of neurons firing simultaneously. Computers are primarily *serial* devices, executing comparatively few operations at a time, such as adding two numbers. But, because computers process millions of operations per second, they achieve fast overall performance. So, while computers process relatively few operations quickly, neurons process copious operations slowly.

Serial and parallel processing each have distinct trade-offs. Parallel processing excels at performing many independent calculations, each requiring a few steps, but is handicapped when many steps are required. This trade-off explains why lightning calculators take several minutes to perform complex calculations. Even though their brains are highly specialized, each step of their calculation depends on the results of the previous one. So, the relative slowness of neurons limits the speed of lightning calculators.

Abstraction

As mentioned, the brain is not a computer, and the mind is not software. This truism cannot be overstated. However, as we now discuss, the relationship between computers and software and the relationship between the brain and mind share surprising commonalities.

Programming early computers, such as the ENIAC, was tedious. Their internal states were altered by manipulating dozens of dials and thousands of plug connections. As early computer pioneers added more memory and capabilities, the manual job of programming these complex machines became exponentially more difficult.

During World War II, the Harvard Mark 1 computer calculated projectile paths for the Navy. Like modern-day computers, the Mark 1 was a binary computer, requiring engineers to transcribe calculations and data to ones and zeros. The settings were specified by punching holes into paper cards or paper tape. A hole in the paper represented a one and the absence of a hole represented a zero.

Mathematician Grace Hopper and her colleagues recorded frequently-used configurations of ones and zeros in notebooks.[159] Having these settings at the ready saved time, but translating all those ones and zeros to paper was still an error-prone process. One wrong setting among thousands could stop the computer cold with no indication of what went wrong. One day, Hopper's settings were correct, but her software malfunctioned. Investigating, she found a moth had made its way into the Mark 1 and shorted a circuit, causing the error. Hopper subsequently referred to any errors within her software as a "bug," a programming term that lives on today.

The Mark 1 outperformed its human counterparts. Nonetheless, Hopper saw room for improvement. Typically, the ones and zeros the operators entered represented a language equivalent, such as "Initial angle of projectile = 33 degrees" or "Distance = 15 nautical miles." This one-to-one relationship between the ones and zeros the computer understood and their language equivalent that humans understood gave Hopper an idea—let the computer do the conversion of language into the ones and zeros. So, rather than operators entering ones and zeros, let them enter words. Her approach to converting human language into machine language became known as "compiling," which forever changed computer science.

Regarding our discussion of the mind, a takeaway from Hopper's insights is that we can abstract large, complex systems into very simple concepts with little or no loss of understanding. However, the tradeoff for using abstraction is that we distance ourselves from the underlying processes. So, entering words instead of numbers is easier for humans, but the abstraction distances us from the configuration of the computer.

This same analogy applies to the brain and mind. By abstracting the brain into schemas and other abstract concepts, we gain insights into the information processes of the brain, but distance ourselves from the underlying neural processes. To understand a system is to know both its abstraction and its concrete implementation. The relationship between the mind and the brain is no different. To fully understand the mind, we must understand the neurons that produce it. And, to fully understand the brain, we must understand schemas and other mental abstractions.

As discussed, the analogies drawn between our parallel-processing minds and serial-processing computers are limited. However, understanding that software merely abstracts a computer's configuration of ones and zeros lends powerful insights into the relationship between computers and their software. Likewise, understanding that the mind merely abstracts the brain sheds light on the relationship between the brain and mind.

Imagine teleporting a modern laptop computer into the late nineteenth century. What would onlookers think as the device powered up? Predating computers and televisions, many would conclude it was from the spiritual world, perhaps delivered from God. Others would turn to science to understand the contraption, deducing that it required electricity and was very much of this world. Using instruments of the day, scientists would eventually discover that the innards of the machine contained silicon that housed tiny, barely perceptible circuits capable of retaining electrical charges. How would they come to understand the organization of billions of electrical bits of information working together to comprise the wondrous machine before them? Without the concept of computer software, how would their limited knowledge of computer hardware lead to a higher understanding of the device? The task before them would seem insurmountable.

Today's scientists attempting to reverse-engineer the brain have an even more daunting task. Understanding the configuration of billions of neurons that organize to form the mind likewise seems insurmountable. Further complicating matters, the complexity of the brain and mind surpasses that of a modern-day computer by several orders of magnitude. Researchers grappling with these questions have their work cut out for them.

Emotional selection portrays the mind as evolving during sleep. Unlike a computer that requires engineers to develop improvements, the mind is self-developing and self-testing. Imagine if a computer could likewise evolve on its own. So, that computer your parents purchased for you as a kid would continue to evolve while sitting on your desk. Its RAM would increase, its graphics and central processors would grow more powerful, and its monitor would grow larger. You would never need to buy a replacement, as it would continually improve its capabilities to meet your needs. Unfortunately, unlike our brain and mind, a computer does not evolve independently.

The brain and mind coexist. Without the structure characterized by the mind, a brain would amount to little more than a jar of neurons. Likewise, the mind abstracts the brain. So, without the brain, there is no mind. Similar analogies exist in computing. A computer without software is as useful as a bag of silica. Software without a computer is equally useless. Together, they form a whole that is greater than the parts. However, once fused, treating them as an indecipherable unit is limiting.[160] When troubleshooting computer malfunctions, technicians determine whether the problem is software-related or hardware-related. If the former, the state of the system is altered. If the latter, then physical parts of the system, such as RAM or chips, must be replaced. The brain sciences face an analogous predicament. The focus of psychology is the mind, while that of neurology and psychiatry is the brain. A challenge for practitioners is determining which mental illnesses are physiological, so best cured with drugs, which are psychological, so best treated with therapy, and which are a combination of physiological and psychological, so best treated with both drugs and therapy.

Neurologist Antonio Damasio appreciates this complexity of the brain, calling it "a system of systems."[161] He proposes that the brain can be studied logically to understand its underpinnings. However, he does not suggest that the brain is merely a rational machine:

> The neurological evidence simply suggests that selective absence of emotion is a problem. Well-targeted and well-deployed emotion seems to be a support system without which the edifice of reason cannot operate properly. These results and their interpretation called into question the idea of dismissing emotion as a luxury or a nuisance or a mere evolutionary vestige. They also made it possible to view emotion as an embodiment of the logic of survival.[162]

Damasio understands the importance of emotion in the inner workings of the mind. When computers someday feel and emote, perhaps the analogy of the brain-as-computer will have fewer caveats.

Unconscious Schemas

Schemas often slip in and out of consciousness. Consider the schemas at work while you drive a car. After starting your car, you pull out of the driveway to run errands. Schemas such as "stay between the white lines" and "signal when turning" move to the forefront of your mind. As you settle into your drive, your mind may wander to your list of daily chores or an upcoming deadline at work. After a few minutes, your attention reverts to the task at hand, and you realize you were "on autopilot" and missed your exit. In those moments, you consciously forgot about driving. However, your unconscious mind ensured that your driving schemas remained active.

As discussed, the schemas that model unconscious thought often conflict. Such conflicts explain the extraordinary case of a woman in the early twentieth century who walked into her doctor's office with an unusual complaint. Every so often, her left hand reached upward and tried to strangle her.[163] To save herself, she fended off her murderous left hand with her right hand, reining it in and forcing it to return to her side. Occasionally, she sat on her left hand to restrain it.

Her left hand operated unconsciously, while her conscious right hand protected her. Why would the goals of her consciousness differ from those of her unconscious? Many specialists concluded she was mentally ill. However, her physician, Kurt Goldstein, arrived at an alternate explanation. Knowing the right side of the brain controls the left hand and that the right brain is more emotional than its left-sided counterpart, Goldstein reasoned that his patient's rational left brain restrained the suicidal tendencies of her more emotional right brain. He concluded she likely had a stroke that severed the corpus callosum—the fibers that connect the left and right brain. Goldstein suggested that, prior to her stroke, the rational thinking of her left brain offset the impulsive thinking of her right brain, but now that her brain's hemispheres were severed, the suicidal urges of her right brain grew unabated.

The woman eventually died from a stroke. An autopsy revealed that her corpus callosum had indeed been damaged by a previous stroke, just as Goldstein hypothesized. This severing split her schema network in two, leaving her with one mental model that wanted to leave this world and the other that wanted to stay in it.

When considering schemas, whether they be unconscious schemas or conscious ones, understanding the relationship between schemas and the neural networks that support them can lead to a deeper understanding of the mind and dreaming.

Discussion

In this appendix, I adopt a systems approach to characterize the mind and its relationship with the brain.[164] Considering the mind as simply an abstraction of the brain leads to powerful insights into the relationship between the brain and mind. These insights are the foundation of the framework that leads to a deeper understanding of dreaming.

To understand the role that dreams play in the daily evolution of the systems of our mind, a general understanding of systems is helpful. In the next appendix, we will review properties of the ubiquitous systems that surround us and comprise us.

A System of Systems

Human beings, viewed as behaving systems, are quite simple. The apparent complexity of our behavior over time is largely a reflection of the complexity of the environment in which we find ourselves.

— Herbert Simon

SYSTEMS SURROUND US. From *Thinking in Systems* by Donella Meadows:

> A school is a system. So is a city, and a factory, and a corporation, and a national economy. An animal is a system. A tree is a system, and a forest is a larger system that encompasses subsystems of trees and animals. The earth is a system. So is the solar system; so is a galaxy. Systems can be embedded in systems, which are embedded in yet other systems.[165]

Systems theory is the study of how parts and processes interact. While the field of systems theory is relatively new, having been formally defined in the mid-twentieth century, its underpinnings are old. Ludwig von Bertalanffy, a founder of general systems theory, outlined how earlier theorists typically thought about the interrelationships between parts of systems. From Ludwig von Bertalanffy's introductory text, *General System Theory*:

> The "Copernican Revolution" was more than [a better way] to calculate the movement of the planets; general relativity more than an explanation of a very small number of recalcitrant phenomena in physics; Darwinism more than a hypothetical answer to zoological problems; it was the changes in the general frame of reference that mattered.[166]

Copernicus' description of the sun as the center of the solar system, rather than the earth, introduced a deeper understanding of the universe and our place in it. Other systems thinkers that predated the field of systems theory include Einstein, whose theory of relativity answered confounding questions about the relationship between space and time, and Darwin, whose theory of natural selection described how species evolve through competition over limited resources.

Designers, engineers, and scientists often create models that describe the organization and internal relationships of systems. While helpful, such models are always limited. As Meadows wrote, "All models, whether mental models or mathematical models, are simplifications of the real world." So, models do not fully capture the complex processes they represent, and that is okay. The intent of systems theory is not to describe a system comprehensively. Instead, systems theory identifies and characterizes important relationships within.

The recognition that systems surround us provides a deeper understanding of our lives. Likewise, we gain insights into emotional selection and its relationship to dreaming by considering the mind as a system.

Schema networks are complex systems that abstract neural processes. In this appendix, we investigate common properties of systems to form a better understanding of emotional selection and the system that is you.

Feedback Cycles

Systems often monitor their performance and make internal adjustments based on this feedback. They then maintain their new state until the next round of feedback and subsequent adjustments. This endless cycle of feedback followed by adjustments is referred to as the "feedback cycle."

As an example of a feedback cycle, consider your home's heating and cooling system. When you set the temperature at 70 degrees Fahrenheit (21 degrees Celsius), your thermostat establishes a "close enough" range. When the feedback indicates the air temperature is below this range, the heat turns on. When the feedback indicates the temperature exceeds this range, the cooling kicks in.

Engineers and consumers routinely "balance" feedback cycles. As seen in our heating and cooling system example, a "Goldilocks range" (not too hot and not too cold) is established.

Psychologists and therapists have observed that people's mental health can depend on feedback cycles. Clinical psychologist Derek Milne explains:

> When things are going well, we talk of a "virtuous cycle," as one success follows another, strengthening our confidence and with it our effectiveness: we can do no wrong, feeling confident and effective. For instance, if we have joined a club for social contact and found someone to talk to, this may encourage a quick return and greater involvement, building to true membership of that group,

> so that it becomes part of one's new identity. When things go less well, we can enter a downward spiral, the so-called "vicious cycle," in which one setback follows another, weakening our confidence and fostering avoidance, self-pity, blaming others, and so on.[167]

Vicious cycles can occur when trauma victims turn to drugs and alcohol for self-medication to ease emotional pain. What appears initially to be a harmless respite from trauma can set in motion a cycle that spirals beyond control, as observed by the authors of *The Anatomy of Addiction*:

> Once a patient begins to self-medicate, a vicious cycle has begun that brings the individual into a whole new ball game. This is because the patient now has concurrent and interrelated problems that feed off of and play into each other. It then becomes not a matter of which came first—the trauma that caused the addiction or the addiction itself—but which disorder to treat first to stop the cycle from spinning even more out of control.[168]

Trauma victims often misinterpret their traumatic experiences as feedback on their behavior, mistakenly blaming themselves for assaults and erroneously believing they could have done or said something to avoid being victimized. Societies often provide additional abhorrent feedback by shaming victims to avoid responsibility. In other words, our mental systems must often cope with inaccurate feedback.

While awake, the continuous feedback that is our life bombards the systems of our mind. During sleep, dreams also provide feedback. While dreaming of being chased down an alley, we may turn and face our attacker. The assailant of our dreams responds, resulting in feedback for our actions. Fortunately, emotional selection uses this feedback to heal damaged schemas and develop new ones to prepare us for the feedback we receive while awake.

Patterns

Dream themes often follow patterns. Themes of teeth falling out, flying, and being accidentally naked in public are common patterns across cultures around the world. As we discussed in Part Two, recognizing these patterns helps categorize and investigate the needs associated with these themes.

The existence of patterns in dreaming should come as no surprise. Both human-made and naturally evolved systems often exhibit remarkably similar patterns. Predators have teeth, beaks, and claws to take down prey. There is no rule that the perimeter of a predator's mouth must be sharp or the tips of their toes must have claws, but these characteristics work, so the pattern evolved.

Emotions, too, follow patterns. For instance, emotions are not amorphous but fall into distinct categories, such as love, anger, lust, and joy. Categorizing emotional patterns helps to identify how we feel and communicate our emotions and feelings with others. Carroll Izard describes:

> In Western cultures (and probably in others as well), people need the category label of *joy* (or its equivalent) to explain the pride of achievement, *sadness* to explain the experience of a life-changing loss, *anger* to explain the frustration of blocked goal responses, and *fear* to explain flight to one another for safety. People need the notion of a pattern of sadness and anger schemas to explain reactions to the terrorist disaster of 9/11/2001, and they need a pattern of fear and anger schemas to explain the effects of the perceived risks of terrorism. To para-

> phrase Bertrand Russell, a discrete emotion category is a convenient way to collect a set of components and characteristics into a useful bundle.

Psychiatry and psychology use patterns to categorize behaviors associated with mental illness. *The Diagnostic and Statistical Manual of Mental Disorders* (DSM) defines mental illness using the vocabulary of patterns, describing a mental disorder as "conceptualized as a clinically significant behavioral or psychological syndrome or pattern that occurs in an individual and that is associated with present distress."[169] As with all patterns, their application has limits. The authors of the DSM have yet to categorize some symptoms, while others fall under multiple diagnoses. Regardless of the discipline, practitioners who apply patterns must accept their limitations.

As we are surrounded by systems, we are also surrounded by the patterns within these systems. The architect Christopher Alexander recognized and applied patterns to describe how people interact with the built environment. He showed that towns and buildings incorporate several patterns, such as shopping streets, nightlife, sacred sites, town halls, and transportation. Rather than being explicitly designed, such architectural patterns often evolve as a community grows from a hamlet to a village, then to a town, and finally to a city.

The patterns within human needs often interplay with the patterns of other systems. Our need for belongingness leads us to form groups and social networks, which are indispensable for meeting countless other needs, such as physiological and safety needs. In turn, these needs drive other patterns, such as those identified by Alexander. For instance, our security needs drive us to build shelter, our social needs drive us to form towns, our cognitive needs drive us to attend school, and so forth. Recognizing the patterns of human needs helps us understand human nature.

Identifying patterns is a crucial step in understanding the parts and processes of various systems, whether they are human-made or the systems that comprise us. When we recognize a pattern, we gain insight into the system it describes.

Self-Organization

In 1952, sixteen years after describing his Turing Machine, Alan Turing published a paper on morphogenesis—the capability of small entities, such as cells, to self-organize into complex structures. At the time, Turing wondered how the cells of a leaf "know" what to do. Cells receive no hierarchical instructions from other cells. Rather, they make "decisions" based on the limited information provided by their neighbors. During his investigation, he found equations that described this behavior. Cells following these "rules" could form complex structures, such as leaves, plants, and organs, without knowledge of their place within the overall structure.

Turing's description of the self-organization of cells revolutionized how researchers consider the complex behavior of groups. One such phenomenon that particularly baffled naturalists was the behavior of slime mold. Slime mold is a ubiquitous, single-celled organism that clusters into colonies. A characteristic of slime mold is that it seemingly disappears during dry, warm weather, reappearing in cool, damp conditions. In actuality, during drier conditions, the single-celled organisms disperse. But under the proper conditions, thousands of cells coalesce into a group that moves as though it were a single organism. Because the cells are identical, the colony lacks digestive, nervous, pulmonary, and other centralized systems. The observation that such colonies perform complex tasks, such as navigating towards food and shelter, all without a brain or overt direction, perplexed researchers.

Prior to Turing's paper, scientists assumed that slime mold colonies followed a hierarchy, with a few "pacemaker" cells giving directives to neighboring cells, such as "go left." The neighbors would pass the signal along to other cells, much like the childhood game of "telephone." This analogy was generally accepted, but one problem stood—no one ever located the pacemaker cells.

Evelyn Fox Keller and Lee Segal of MIT wondered whether the principles of self-organization explained the behavior of single-celled organisms such as slime mold. They developed a set of equations that described the behavior of the slime mold cells without pacemakers. In 2000, Toshiyuki Nagasaki published that he had dramatically demonstrated this behavior by training slime mold to navigate a maze. In his experiment, he placed the slime mold in a simple maze with four possible routes to two food sources. After "learning" the maze, the slime mold found the shortest routes, stretching its amoeba-like "body" between the two food sources, all without eyes, a nervous system, or a brain.[170]

This will be an utterly unflattering analogy, but our schemas share similarities with slime mold cells. A schema's understanding of the outside world is also limited to the information relayed by its neighbors. However, there is a significant difference between the cells of slime mold and the neurons that comprise our brains and produce our schemas—neurons form far more connections with other neurons. In short, they have more neighbors. So, neurons and schemas receive and distribute more information than a simple slime mold cell, which explains why a schema network can complete more complicated tasks than merely navigating a maze.

Self-organization describes simple rules that define the outwardly complex behaviors of many systems, from ant colonies and beehives to cities.[171] It also helps us explore how schemas self-organize to accommodate new information. As discussed, schemas improve themselves through dreams. In other words, schemas use dreams to self-organize.

Local Optimization

As we enter new friendships, change jobs, or find ourselves in new situations, our schemas recognize their limitations and set to work self-organizing to improve their ability to meet our needs. Will our schemas reach their full potential? Sometimes yes, sometimes no. A challenge facing schemas is what engineers refer to as *local optimization*.

As an example of local optimization, consider the hypothetical situation of living in a scenic valley at the foot of a dam. One morning, you wake to blaring sirens warning of an imminent dam failure. You gather your loved ones and head for higher ground. Two hills are nearby. The nearest one is shorter than the one in the distance. Fearing the dam will burst before you reach the taller hill, you and your family scramble up the shorter one. You reached higher ground, but it is local.

The dam bursts, and the floodwaters rise. Your modest hill is now an island. As your island shrinks in size, you look longingly at the distant, higher ground. Should you chance the swim before the water puts more distance between you and it? Or should you stay on your suboptimal hill in the hope that the high-water mark never reaches your feet? Your locally optimized hill has kept you dry, but it has its limitations.

Local optimizations can vex both engineers and the mind. To reach a "higher" optimization, systems must travel through "lower" terrain. For human-made systems, engineers typically seek more optimal solutions while the system is offline. Once found, engineers implement the improved optimization, and the system is brought back online. The mind exploits sleep for the same purpose—to seek more optimal solutions for its schemas. Once found, we awake with a more adaptive view of the world and our place in it.

In his insightful book, *The Design of Everyday Things*, Don Norman describes the built environment as a product of design. He celebrates designs that work and highlights those that are suboptimal. His insights on the latter led to his moniker being placed on poorly designed doors with no indicators of whether they should be pushed or pulled. Referred to as "Norman doors," these doors have a handle suitable for pulling on both their push side and the pull side, unnecessarily confusing the user. When our first attempt to open a Norman door fails, we typically do not deem the door unusable. Instead, we try the opposite direction. (Or, if we are Don Norman, we write a book imploring architects to develop more user-friendly designs.)

Emotional selection works to improve our ability to meet our needs, yet sometimes it arrives at suboptimal solutions that are the equivalent of a Norman door or a local hill at the foot of a breached dam. A sign that schemas operate at a local optimization is an unhealthy habit that is hard to break, such as self-medicating or adhering to old routines that no longer fit our lives. As we discussed in Part Two, recognizing locally optimized schemas is a step toward developing better approaches to meet our needs, which clears a path for emotional selection to help your schemas "swim for higher ground" while you sleep.

Discussion

At our most basic level, humans are a system of cells. While these cells self-organize into a respiratory system, circulatory system, and other systems of the body, such descriptions fail to capture our complexity. Rather, only a holistic description that includes interactions with our loved ones, peers, and surroundings approaches a satisfying description of what it means to be human.

Systems theory provides a structure for describing and discussing characteristics shared among outwardly dissimilar systems. However, a musician can compose and perform beautiful music without having formally studied music theory. Likewise, you do not need to be an expert in systems theory to apply its principles. For instance, we can talk about the feedback of a heating system in the same context as the feedback of the mind. The two are outwardly dissimilar, yet share common principles.

While the idea that dreams modify and test schemas during sleep may appear radical, the simple concepts of modification and testing are as old as evolution itself. In the next appendix, we consider the various tests that surround us and explore their surprising similarities to many of our dreams.

Modify, Test, Repeat

Good tests kill flawed theories; we remain alive to guess again.

— Karl Popper

SOME SYSTEMS are easier to test than others. "Black box" systems are so named because their inner workings are hidden. Engineers testing black box systems are restricted to monitoring inputs and outputs. "Clear box" systems are easier to test because their internals are observable, so engineers can better monitor when the system goes awry. "Gray box" systems, as the name implies, fall between clear box and black box, describing systems with partially observable internals.

Prior to the mid-twentieth century, the brain was a black box. Researchers understood that neurons interconnected to form the brain, but could not directly observe their processes. This changed with the advent of EEGs and fMRIs. Thanks to these brain imaging tools, the brain can now be classified as a gray box. However, brain imaging only shows regions of brain activity; they cannot yet detect what individual neurons and schemas are "thinking." Someday, the brain may become a clear box. Until that day, our ability to intuit its subsystems is limited.

Regardless of the box categorization, systems are subjected to a variety of tests. "Stress" testing evaluates performance limits—think NASA dropping a spacecraft onto concrete to simulate a hard landing. "Unit" tests evaluate whether an input produces an expected output. "Fuzz" testing evaluates whether designs can withstand a comprehensive range of random inputs. And "regression" testing evaluates whether prior changes to a system introduced errors.[172]

Dreams that test our schemas are analogous to the tests engineers use to evaluate systems. Dreams that push schema modifications to their limits with stressful dream scenarios are stress tests, dreams that ensure clusters of schemas perform as expected are unit tests, and dreams that ensure changes to our schemas do not introduce maladaptive conflicts are regression tests.

To understand dreaming, a comprehensive review of testing methodologies is unnecessary. However, an appreciation of the role testing plays in human-made systems, natural systems, and the systems of our mind and brain is helpful.

Nature's Tests

From *The Origin of Species*:

> It may be said that natural selection is daily and hourly scrutinizing, throughout the world, every variation, even the slightest; rejecting that which is bad, preserving and adding up all that is good; silently and insensibly working, whenever and wherever opportunity offers, at the improvement of each organic being in relation to its organic and inorganic conditions of life. We see nothing of these slow changes in progress, until the hand of time has marked the long lapses of ages, and then so imperfect is our view into long past geological ages, that we only see that the forms of life are now different from what they formerly were.

Darwin explained how species evolve without requiring the guidance of a higher being. Instead, each generation varies in subtle yet important ways. A bird couple may produce chicks with varying coloration, wingspan, or beak lengths. Primate parents may produce children with differing bone structure, muscle composition, and cranial capacity. These variations result in some offspring being better adapted than others. The struggle for survival then ensues, with those less fit being more likely to be culled from the gene pool, leaving the more adaptive offspring to pass their traits along to the next generation.

Nature is not the only decider of which traits to keep. Humans have co-opted natural selection to modify species for our benefit. Breeders select plants and animals with traits that we, not nature, find beneficial. By using a combination of crossbreeding and selection, we created many self-serving variations, such as high-yield crop species and ornamental roses. We domesticated dogs from wolves and cultivated dairy cows with immense udders—much larger than required for nourishing their offspring.

On this topic, Darwin remarks:

> One of the most remarkable features in our domesticated races is that we see in them adaptation, not indeed to the animal's or plant's own good, but to man's use or fancy... We cannot suppose that all the breeds were suddenly produced as perfect and as useful as we now see them; indeed, in several cases, we know that this has not been their history. The key is man's power of accumulative selection: nature gives successive variations; man adds them up in certain directions useful to him. In this sense, he may be said to make for himself useful breeds.

Throughout the history of life on Earth, dating back to the dinosaurs and beyond, most species that came into existence are now extinct. This observation does not imply that natural selection is somehow flawed. Natural selection contributes to the perpetuity of life. As one species becomes extinct, another, more adapted species takes its place. So, while natural selection continually adapts a species, its perpetuity is not guaranteed. This same principle applies to our schemas. Schemas that become maladaptive must adapt to remain within the schema network. Otherwise, they become extinct, replaced by more adaptive schemas.

Trial-and-Error

The trial-and-error approach to design introduces changes to a system, called *trials*, that are somewhat random. A trial that fails to improve performance is said to be an "error" and is rejected. Those that improve performance are adopted. Regardless of the result, the cycle is repeated until design goals are met.

Trial-and-error appears ad hoc, yet produces impressive results. When developing the first light bulb, Thomas Edison did not calculate elaborate formulas to determine which material constituted an ideal filament. Rather, he performed random trials until he found that carbon produced light when an electric current passed through it. Natural selection is another example of trial-and-error—generational variations being the trials and offspring that die before procreating being the errors.

Historically, medical practitioners have used trial-and-error with natural products, such as herbs, in various combinations when seeking cures. German folk medicine discovered that a sometimes cure for stomach ailments is drinking the broth that accumulates at the top of a vat of sauerkraut, now known to contain the penicillium mold. Similarly, tea drinkers found that tea made from white willow bark, now known to contain aspirin, relieved common aches.

Trial-and-error has its limitations. Imagine if automobile designers applied the literal lessons of natural selection. Every automobile rolling off an assembly line would contain unique, subtle variations. The diameter of the wheels, characteristics of the steering, power of the engine, and countless other characteristics would vary slightly. Engineers would then label automobiles as errors if they wrecked, stopped running, injured their occupants, or performed poorly in other ways. The changes in cars still functioning would be incorporated into the next generation. Such an approach is obviously absurd, as it would not be cost-effective nor tolerated by the public. So, rather than taking their cue from natural selection, engineers perform their testing in controlled environments.

Iterative Design

When designing systems, instead of introducing random variations, designers often intelligently guide modifications. This process avoids obvious, ineffective changes, resulting in fewer failed tests than trial-and-error.

The iterative design process typically begins in the development phase, in which engineers and designers modify a system. The testing phase follows, during which modifications are evaluated. Those that pass testing are integrated into the system. Modifications that fail testing are rejected or further modified and tested. Regardless of the outcomes, the process returns to the development phase and begins anew.

NASA demonstrated a clear example of iterative design on its Mars rover, Spirit. On January 4, 2004, the rover began its exploration of the red planet. In communication with NASA, Spirit received instructions from scientists and began its first tentative steps. Unfortunately, after two weeks of exploration, a bug in Spirit's software caused its flash memory to fill up. Engineers had programmed Spirit to reset itself when it detected problems, but resetting did not clear the memory, so this limited ability to self-modify failed. Unable to proceed with its mission, Spirit put itself in sleep mode.

NASA engineers set to work on the problem here on Earth. Using a surrogate rover, they continually modified and tested. Eventually, they repaired the software bug and radioed the changes through space. Spirit received and installed the software changes. The rover then "awakened," better adapted to its environment.

The development and testing the NASA engineers performed is surprisingly analogous to emotional selection. Before incorporating changes to the rover on Mars, modifications were iteratively implemented and tested in a controlled environment on Earth while the rover "slept." When deemed successful, engineers incorporated the changes, and the rover awakened more adapted to begin its day.

Natural Selection Versus Iterative Design

It is tempting to declare that iterative design is the superior design process compared to trial-and-error because it involves a designer who acts deliberately, rather than relying on variation guided by simple rules. However, trial-and-error methods routinely out-design humans. One such design method used by the scientific community is *genetic algorithms*.

Multivariate problems are problems with multiple inputs. These problems can vex engineers and scientists because they have several solutions, each with distinct trade-offs. One approach to solving multivariate problems is to test every combination of inputs. The challenge is that the number of combinations can be so large that even supercomputers cannot calculate every possibility in a reasonable amount of time.

Genetic algorithms evolve toward a solution without having to try every combination of inputs, making them ideal for solving multivariate problems. Based on the success or failure of an outcome, the algorithm adjusts and reapplies input parameters. Like natural selection, the more variations tried, the more optimal the evolution becomes.

The elegant solutions generated by genetic algorithms often surprise the humans overseeing them. Scientists used a genetic algorithm to determine the optimal orbit for telecommunication satellite coverage.[173] High-orbit satellites give better ground coverage than low-orbit satellites, but are more expensive. Low-orbit satellites are cheaper, but sometimes lose line-of-sight views, resulting in blackouts. The challenge is to develop orbits in which line-of-sight is preserved at the lowest possible cost.

Symmetric designs for the orbits of three to five satellites have existed for years. When tasked with designing more optimal orbits, a genetic algorithm produced an unconventional solution. Rather than symmetric orbits, the genetic algorithm produced asymmetrical orbits, with satellites alternating between being very close to each other and orbiting a great distance apart. This unusual configuration reduced both the number of blackouts and the typical duration of each blackout. Researcher William Crossley noted that "engineers with years of aerospace experience were surprised by the higher performance offered by the unconventional design."

A genetic algorithm also produced a surprising truss design for use on the space station and satellites.[174] The design created by the algorithm minimized the transfer of damaging vibrations along the truss and was no more difficult to build than the conventional designs of engineers. Ironically, the researchers commented that the design resembled a human leg bone. The deployment of the truss was an unmitigated success. Yet, as author Charles Petit noted, "No intelligence made the designs; they just evolved."[175]

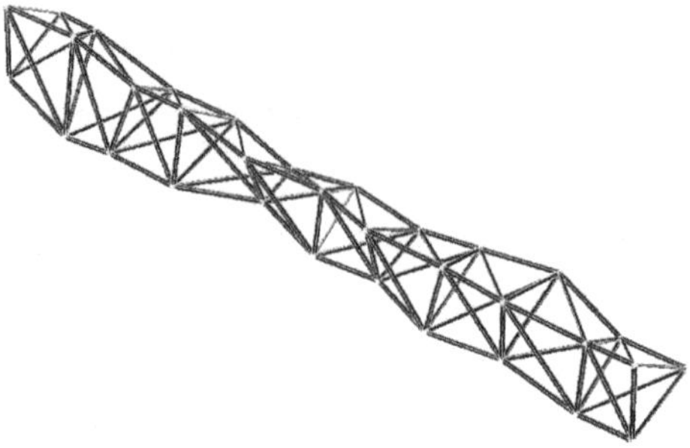

Space station truss. Image courtesy of Andy Keane

Humans may like to think that, given enough time, we will develop the technology and understanding to surpass the designs of nature. Genetic algorithms have thrown a wrench in this reasoning. Humans are more intelligent than algorithms, so our design processes require fewer iterations. However, given enough time, the trial-and-error processes of genetic algorithms and natural selection are often superior to those of humans, suggesting that, as human-made designs become increasingly complex, we will likely turn to natural selection-like design methods, such as genetic algorithms, to solve our most difficult problems.

As observed with natural selection and genetic algorithms, the products of unguided design processes can be surprisingly sophisticated. Emotional selection is also an unguided process. We do not give our mind instructions on what to dream about as we doze. Instead, as we sleep, emotional selection independently chooses which schemas to modify and test with dreams.

Scenario Tests

Scenario tests mimic real-world situations that a system may encounter. Engineers conduct scenario tests in controlled environments where they evaluate the system's performance. These tests can be gentle, mildly testing a system to ensure that it works as intended. They can also be extreme, taxing the system to its breaking point or beyond. Although the concept of scenario tests is simple, designing them can be tricky, as complex systems often contain thousands of components, each with multiple connections.

In the first half of the twentieth century, automobile engineers prioritized performance over safety, leading to countless unnecessary fatalities. In the 1950s, priorities changed as the general public began questioning the high number of road casualties. Crashworthiness tests are scenario tests born out of these concerns.

Considering dreams as scenario tests explains why dreams can be both tame and taxing. Sometimes emotional selection creates gentle dream tests to ensure schemas work as intended. Other times, it creates extreme, bizarre dreams that tax our schemas to ensure they can cope with trauma while awake. Regardless of their intensity, the consideration of dreams as scenario tests for our schemas can lend insights into the role dreams play in evolving our minds.

Discussion

Tests are an integral part of life. Engineers employ tests to develop the machinery, devices, and processes we utilize every day; natural selection uses tests to evolve species; and emotional selection uses tests to evolve your schemas while you sleep.

This appendix concludes our review of the diverse background of emotional selection. We began with a discussion of the remarkable ways in which our schemas generate perception, including how schemas often distort our perceptions to enhance our ability to adapt to our surroundings. We then pivoted to the ubiquity of systems, exploring those that surround us and comprise us. From there, we investigated the systems of the mind, considering the mind as an abstraction of the brain. We concluded with an overview of how virtually all systems undergo testing, including how dreams serve as scenario tests of our system of schemas while we sleep.

There is certainly more support for emotional selection than I could fit within the pages of this book. I encourage the reader to reflect on the role of systems, testing, and dreams during your daily life in search of even more evidence. It is truly remarkable the cross-over between these outwardly unrelated processes that surround us and comprise us.

Afterword

Dear Reader,

As these pages draw to a close, I want to extend my heartfelt gratitude for accompanying me on this journey. Your time and engagement throughout this book have been as important as the words within.

But our journey doesn't end here. In fact, in many ways, it's just beginning, and you, as a reader, play a crucial role in that next phase.

In today's digital world, your opinions are more valuable than ever. If this book has provided valuable insights or sparked new thoughts, I humbly ask you to consider sharing your experience. Whether you have a star rating in mind or detailed reflections, posting your recommendations and feedback to your favorite literary and social media sites, such as Amazon, Goodreads, Facebook, Instagram, TikTok, and X, will help this book reach a wider audience, foster a broader conversation, and earn you my eternal gratitude.

If you wish to stay current with emotional selection, please visit www.emotionalselection.com, where you can subscribe to our mailing list, join a forum, discuss your dreams with others, and more.

Sweet dreams!
Richard Coutts

Acknowledgments

THERE is no exaggeration in my saying that, if not for the generous support of family, friends, acquaintances, and kind strangers, this book would not exist. I owe a debt to you all. These remarkable souls, roughly in chronological order, include my beta readers Heather Holub, Kelley Banks, John McConnell, my sister Ellen Curran, my mom Bernice Coutts, and the Somerquillians (Danielle Andrews, Stephanie Fitzgerald, Rob Van Dyke, Fabienne Keck, and others). I am also indebted to Barbara Murphy, formally of the MIT Press, who sent me down the publishing path. To my mentor, Ed Pace-Schott, whose deep knowledge and boundless curiosity are matched only by his generosity. To Steph Isbell, who turned my hen scratching into journal articles. To Roz Cartwright, who pushed me to broaden my research. To the team at Beth Israel Deaconess Center who helped me gel my ideas into a hypothesis—these brilliant and kind folks include Ed, Bob Stickgold, David Kahn, and Erin Wamsley. To my study participants, who shared their personal dreams, which contributed profoundly to this book and to my research. To Ilya Blum, the mathematical mind who guided my experiments. To David Streiner, who shepherded me through logistic regression with patience and humor. To Tore Nielsen, who taught me more than he realizes about writing for peer-reviewed journals. And finally, to Melanie and Emma, who sacrificed more than anyone for this book, including yours truly. Melanie and Emma, you finally have your husband and dad back, whether you still want him or not.

About the Author

RICHARD holds graduate degrees from Cornell University and Northeastern University. He was born and raised in Dedham, Massachusetts, lived in nearby Cambridge, and now resides in beautiful Halifax, Nova Scotia. When not researching dreaming or relaxing with his wife Melanie and daughter Emma, he can be found running the streets and trails of Nova Scotia, or on his couch, tapping on his laptop.

Notes

1. Measures of the continuity hypothesis (which we will discuss in Chapter Three), such as dating before sleeping, having a crush on someone, and desiring a relationship, did not confound results. For more information, see Coutts, 2023.

2. See Coutts, 2008.

3. The neural network that supports memory is a subset of the neural network that supports a schema network. For example, neurologist Jay McClelland proposed that the brain exploited its neural connections to associate new learning with known concepts (McClelland and David E Rumelhart, 1981). He theorized that as the brain formed generalized concepts, such as "red" or "friend," it simply connected new memories to them as it learned.

   McClelland demonstrated his ideas with a simplified example of the mind remembering the characters and plot of West Side Story. Before watching the performance, our minds contain generalized concepts, such as "married," "bookie," and "college." Introduced to characters, such as "Lance," our minds simply associate the new character with the existing generalizations by forming connections to them. Since each neuron has up to 10,000 connections, and because associations require only one connection, the approach is very efficient.

   Schema networks would exploit physical connections as described by McClelland, but also combine multiple physical connections to form virtual connections.

4. Notable exceptions are people with narcolepsy, who begin sleep almost immediately with REM sleep and have involuntary bouts of REM sleep while awake. Also, while sleep typically terminates during REM sleep for healthy sleepers (unless interrupted), healthy dreamers also spontaneously awake from non-REM sleep, though less frequently.

5. See E. Dorus, W. Dorus, and Rechtschaffen, 1971.

6. See R. D. Cartwright, 1977, p. 129.

7. See Verdone, 1965.

8. See R. D. Cartwright, 1977, p. 20.

9. See R. D. Cartwright, 1977, p. 23.

10. See R. D. Cartwright, 1977, p. 24.

11. See Meddis, Pearson, and Langford, 1973.
12. See Karni et al., 1994.
13. See Kuriyama, Robert Stickgold, and M. P. Walker, 2004.
14. See Wagner et al., 2004.
15. See Winerman, 2006.
16. See Xie et al., 2013.
17. See Landrigan et al., 2004.
18. See Barger et al., 2005.
19. The National Sleep Foundation's 2005 poll is available at www.sleepfoundation.org.
20. The total of 140,000 sleep cycles was calculated as 5 cycles per night over 77 years.
21. See Maslow, 1954, p. 38.
22. See Maslow, 1954, p. 45.
23. See Goble, 1970.
24. See Coutts, 2020. For clarity, these dream(s) were corrected for punctuation, spelling, and grammar in this book as necessary.
25. See Delaney, 1997, p. 152.
26. See Delaney, 1997, p. 151.
27. See C. S. Hall and Nordby, 1972.
28. See Foulkes and Rechtschaffen, 1964.
29. See Bokert, 1968.
30. See Wood et al., 1992; and Valli et al., 2006.
31. See Knudson, 2001.
32. See Coutts, 2010.
33. See Domhoff and Schneider, 2023.

34. See Domhoff and Schneider, 2023.

35. See Coutts, 2015.

36. See Baumeister, Catanese, and Vohs, 2001.

37. See R. D. Cartwright, 1991 and R. D. Cartwright, 1996. The forty depressed participants comprised twenty men and twenty women.

38. See *What Are Dreams* 2009.

39. See Bernstein, 1985.

40. See Bernstein, 1985, p. 68.

41. See Hartmann, 1996.

42. See Stoddard, Chedekel, and Shakun, 1996.

43. See Wilmer, 1996.

44. See Hartmann, 1996 and Stoddard, Chedekel, and Shakun, 1996.

45. See Maybruck, 1991.

46. See M. Rosen and Sutton, 2013.

47. The lucidity of our perceptions during dreams was also recognized in LaBerge and Rheingold, 1990, p. 102-07.

48. See *What Are Dreams* 2009.

49. See Green, 1968, p. 58.

50. See Green, 1968, p. 48.

51. See Weisberg, 2006.

52. See Berns, 2013.

53. See Jeremy Taylor, 2009.

54. See Morin, 2015.

55. See McCurdy, 1946.

56. See Rycroft, 1979.

57. See Carroll and Prickett, 2008, Genesis, 41:1-57.

58. See Freud, 1983.

59. See Freud, 1983.

60. See Van de Castle, 1994.

61. See Robert, 1886.

62. The term "practice play" is borrowed from naturalists who observe similar behavior in animals. A cat may "play" with a wounded mouse to rehearse for hunting future prey, using exaggerated pouncing and stalking movements. See Fagan, 1976.

63. See Winson, 1990.

64. See Revonsuo, 2000.

65. See Revonsuo, Tuominen, and Valli, 2015.

66. See Hobson and McCarley, 1977.

67. That seagull's trill when a dog is in their vicinity is based on my personal observations while enjoying Boston's waterfront during lunchtime.

68. See Baron-Cohen, 1997.

69. See Darwin, 1873.

70. See R. Cartwright and Lamberg, 1992.

71. Cartwright and Lamberg hypothesized that dreaming better adapts us to our waking environment by aiding the mind in finding the proper balance between such qualities.

72. See Garfield, 2001.

73. See J. Taylor, 1992, p. 195-6.

74. Researchers describe dreamers who perform well during a dream theme as having "mastered" the dream. A dreamer who embraces a pleasant dream of flying or having sex would be said to have mastered a dream. Mastery also applies to unpleasant dreams. Themes of equipment malfunctioning or being accidentally naked in public will be unpleasant for everyone. Dreamers who take the scenario

of the dream in stride, believing that it did not reflect deeper flaws in their character, are said to have mastered the dream, despite its unpleasant theme.

75. See Muller, 1996.
76. See Maslow, 1943.
77. See Beck, 1976.
78. See Young, Klosko, and Weishaar, 2003.
79. See Delaney, 1997.
80. See J. Taylor, 1992, p. 178-9.
81. See Spurr, 2001, 44.
82. See Spurr, 2001, 120.
83. See Garfield, 2001, p. 208.
84. See Siegel, 1996.
85. See Garfield, 2001, p. 228.
86. See Garfield, 2001, p. 230.
87. See Garfield, 2001, p. 93, and Delaney, 1997, p. 156.
88. See Delaney, 1997, p. 72.
89. See Delaney, 1997, p. 111.
90. See Garfield, 2001, p. 192.
91. See Starr, 2001, p. 86.
92. See Garfield, 2001, p. 203.
93. See J. Katz, 1994, p. 56.
94. See Delaney, 1997, p. 69.
95. See Delaney, 1997, p. 99.
96. See R. Cartwright and Lamberg, 1992, p. 24.

97. See R. Cartwright and Lamberg, 1992.

98. See Cacioppo, Hawkley, and John, 2007.

99. See Averill, Long, and James, 2003.

100. See Calvin Springer Hall, 1963.

101. See Emanuel et al., 2000.

102. See Domhoff and Schneider, 2023—Chuck: a physical scientist: #23 (02/20/92).

103. See Domhoff and Schneider, 2023—Barb Sanders #2318 (12/11/92).

104. See Murrell, Meeks, and J. Walker, 1991.

105. See Arndt and Goldenberg, 2002 and Baumeister, Campbell, et al., 2003.

106. While self-esteem encompasses our need to love ourselves, it must not be confused with narcissism. The self-admiration that accompanies narcissism is an unhealthy coping mechanism for deep insecurities. Narcissists often have mediocre intelligence yet view themselves as brilliant or are morally bankrupt yet see themselves as virtuous. Like the mythological Narcissus who fell in love with his own reflection, they may love a singular quality about themselves at the expense of developing a truly lovable persona.

107. See Garfield, 2001, p. 153-69, and Delaney, 1997, p. 151.

108. See Garfield, 2001, p. 169, and Delaney, 1997, p. 29.

109. See R. Cartwright and Lamberg, 1992, p. 79.

110. See Garfield, 2001, p. 176, and Gibson, 1969, p.51.

111. Dreams of being able to fix a car so it "runs beautifully" are reported in Delaney, 1997, p. 31.

112. Dreams of abused animals and children are reported in Delaney, 1997, p. 87-88.

113. See R. Cartwright and Lamberg, 1992, p. 125.

114. See Siegel, 1996, p. 172.

115. See Maybruck, 1991, p. 78.

116. See Harlow, 1950.

117. Sagan continues to say in the dream that he developed alexia when reading the future and consequently could not discern it. This is inconsequential because his mind witnessed the positive response of his perceptions upon making the discovery and did not need to continue the dream by fabricating the future (see Sagan, 1977, p. 153.

118. See Tommaso, 2008.

119. See Chang, C. Chen, and Huang, 2008.

120. See Garfield, 2001, p. 153-69, and Delaney, 1997, p. 151.

121. See Domhoff and Schneider, 2023—Barb Sanders, #1060 (09/15/88).

122. Beauty is not limited to what we sense. Beauty can also be found in what we think. English mathematician and philosopher Alfred North Whitehead proposed that patterns play an integral role in determining art: "Art is the imposing of a pattern on experience, and our aesthetic enjoyment is recognition of the pattern." In the language of schemas, our schemas harbor patterns that continuously decode events in our life. When these patterns describe beauty, wonder consumes us. This beauty can take many forms. It can be a painting on a wall, a piece of music, a dance posture, a mathematical formula, or the grace of an athlete. Beauty surrounds us, and our schemas continually seek it out to meet our aesthetic needs.

123. See Maslow, 1943, p. 382–383.

124. See Garfield, 2001, p. 153-69, and Delaney, 1997, p. 151.

125. See *Dreamtime* 2008.

126. See Maslow, 1971.

127. See Maslow, 1969, p. 270.

128. See Domhoff and Schneider, 2023—Barb Sanders: #1676 (02/27/90).

129. See Sharot, 2011.

130. See Beckman, 1970.

131. See Schoenberger, 1988.

132. See Snyder and Swann, 1978.

133. See Sullivan, 1953.

134. Firestone, 1985.

135. To test his theory, Tononi and fellow researchers had eleven volunteers play a simple video game that involved reaching for objects with a mouse-controlled cursor. Unbeknownst to the participants, the game sometimes introduced a slight bias to the cursor's motion, forcing the participants to adjust their movements. Half the group slept with their heads attached to sleep monitors between sessions while the others remained awake. Monitoring equipment revealed that regions of the sleepers' brains that learned to compensate for the bias while awake had the largest slow-wave activity during sleep. Among the participants who slept, those who recorded more slow-wave activity in the respective part of the brain performed better the following day.

136. See W. Dement et al., 2014.

137. See W. C. Dement, 1974.

138. See Miller, 2007.

139. See Wamsley and R. Stickgold, 2011.

140. Researchers draw distinctions between mental activity during sleep. "Apex dreams" and "non-apex dreams" collectively are what are traditionally considered as dreams. During non-REM sleep, the mental activity is more thought-like then dream-like, so researchers use the terms "cognitive activity" and "cognitive processes." Drawing such distinctions are beyond the scope of this book. For more information, see Nielsen, 2000.

141. Robert Firestone refers to this distortion of the child-parent relationship model as the "fantasy bond," observing that all such relationships are distorted to varying degrees (Firestone, 1985).

142. Schemas with biases against adaptation are referred to as modes by Young (see Young, Klosko, and Weishaar, 2003). These modes often result in *confirmation bias*, which is when schemas assimilate information that supports biases while rejecting those that challenge them.

143. This "town hall" meeting of schemas agrees with the description of schemas as being active processes that recognize the information they store (see David E. Rumelhart, 2017).

144. This description of spindles as communication among schemas in the fashion of a town hall meeting is highly speculative. I offer this hypothesis merely for consideration.

145. See Ekman, 1999.

146. See Izard, 2007.

147. See J. B. Rosen, 2004.

148. See Schultz, Izard, and Bear, 2004.

149. See Izard et al., 2015.

150. See Damasio, 1999.

151. Researchers that consider the brain and mind as a single entity, or who describe the brain as producing the mind, often use the terms "brain-mind" or "mind-brain" to describe the brain and mind as a bound whole.

152. See Hampden-Turner, 1981.

153. See Turing et al., 1936.

154. The advent of computers also contributed to the new field of *cognitive science*, which studies include the processes of voice, facial, and pattern recognition. While AI researchers focus on creating intelligent machines, cognitive scientists strive to understand the underlying processes that give rise to human intelligence.

155. See Crevier, 1993.

156. See Crevier, 1993, p. 109.

157. See *Greetings People of Earth* 2023.

158. See Ramachandran, Blakeslee, and Dolan, 1998, p. 197.

159. See Harford, 2017.

160. Some researchers contend that viewing the brain and mind as separate entities is counterproductive, maintaining that "the brain produces the mind." Extremists in this camp assert that it matters little whether mental illness is treated as a physiological problem or a mental problem, as long as the symptoms are alleviated.

Fortunately, most scientists regard this reductionism, believing that differentiating between the brain and mind is critical to a deeper understanding of both.

161. See Damasio, 1999, p. 331.

162. See Damasio, 1999, p. 42.

163. See Ramachandran, Blakeslee, and Dolan, 1998, p. 12.

164. This definition intentionally omits the role of consciousness in understanding the mind. Cognition describes the ability to think, reason, and remember, and is often referred to as the "easy" question of the mind. Consciousness is the awareness that you exist. Explaining consciousness is considered the "hard" question. For more information, see Chalmers, 2002.

165. See Meadows, 2008.

166. See Bertalanffy, 1968, p. 99.

167. See Atton, 1985.

168. See M. Khaleghi and K. Khaleghi, 2011.

169. See American Psychiatric Association, Association, et al., 1994.

170. See Johnson, 2002.

171. See Meadows, 2008.

172. Other types of testing include smoke testing, acceptance testing, and compatibility testing. For more information, see *Software testing* 2023.

173. See Williams, Crossley, and Lang, 2001.

174. See Keane, 1996.

175. See Petit, 1998.

Bibliography

American Psychiatric Association, AP, American Psychiatric Association, et al. (1994). *Diagnostic and statistical manual of mental disorders: DSM-IV*. Vol. 4. American psychiatric association Washington, DC.

Arndt, Jamie and Jamie Goldenberg (Jan. 2002). *From threat to sweat: The role of physiological arousal in the motivation to maintain self-esteem*, pp. 43–69. ISBN: 1-55798-883-8. DOI: 10.1037/10448-002.

Atton, Hilary (1985). "The psychology of retirement". In: *British Journal of Occupational Therapy* 48.12, pp. 375–378.

Averill, C., R. Long, and R. James (2003). "Solitude: An Exploration of Benefits of Being Alone". In: *Journal for the Theory of Social Behaviour* 33.1, pp. 21–44.

Barger, Laura K et al. (2005). "Extended work shifts and the risk of motor vehicle crashes among interns". In: *New England Journal of Medicine* 352.2, pp. 125–134.

Baron-Cohen, S. (1997). *Mindblindness: An Essay on Autism and Theory of Mind.* MIT Press, pp. 13–14.

Baumeister, Roy F, Jennifer D Campbell, et al. (2003). "Does high self-esteem cause better performance, interpersonal success, happiness, or healthier lifestyles?" In: *Psychological science in the public interest* 4.1, pp. 1–44.

Baumeister, Roy F, Kathleen R Catanese, and Kathleen D Vohs (2001). "Is there a gender difference in strength of sex drive? Theoretical views, conceptual distinctions, and a review of relevant evidence". In: *Personality and social psychology review* 5.3, pp. 242–273.

Beck, A. T. (1976). *Cognitive therapy and the emotional disorders.* International Universities Press.

Beckman, L. (1970). “Effects of Students’ Performance on Teachers’ and Observers’ Attributions of Causality”. In: *Journal of Educational Psychology* 61, pp. 76–82.

Berns, Gregory (2013). “Dogs are people, too”. In: *The New York Times* 5.

Bernstein, Isidor (1985). “Children’s Dreams: Longitudinal Studies. By David Foulkes”. In: *Journal of the American Psychoanalytic Association* 33.1, pp. 237–241.

Bertalanffy, Ludwig von (1968). *General system theory: Foundations, development, applications*. G. Braziller.

Bokert, E. G. (1968). “The effects of thirst and a related verbal stimulus on dream reports”. In: *Dissertation Abstracts International* 28.11-B.

Cacioppo, L., C. Hawkley, and T. John (2007). “Aging and Loneliness: Downhill Quickly”. In: *Current Directions in Psychological Science* 16.4, pp. 187–191.

Carroll, Robert and Stephen Prickett (2008). *The Bible: Authorized King James Version*. Oxford Paperbacks.

Cartwright, Rosalind and Lynne Lamberg (1992). *Crisis dreaming: Using your dreams to solve your problems*. HarperCollins.

Cartwright, Rosalind D (1977). *Night life: Explorations in dreaming*. Prentice-Hall.

— (1991). “Dreams that work: The relation of dream incorporation to adaptation to stressful events”. In: *Dreaming* 1.1, p. 3.

— (1996). “Dreams and adaptation to divorce”. In: *Trauma and dreams*, pp. 179–185.

Chalmers, David John (2002). *Philosophy of mind: Classical and contemporary readings*.

Chang, M., C. Chen, and K. Huang (2008). “Effects of music therapy on psychological health of women during pregnancy”. In: *Journal of Clinical Nursing* 17, pp. 2580–2587.

Coutts, Richard (2008). "Dreams as modifiers and tests of mental schemas: an emotional selection hypothesis". In: *Psychological reports* 102.2, pp. 561–574. URL: https://doi.org/10.2466/pr0.102.2.561-574.

— (2010). "A pilot study for the analysis of dream reports using Maslow's need categories: An extension to the emotional selection hypothesis". In: *Psychological reports* 107.2, pp. 659–673.

— (2015). "Variation in the Frequency of Relationship Characters in the Dream Reports of Singles: A Survey of 15,657 Visitors to an Online Dating Website". In: *Comprehensive Psychology* 4. URL: https://doi.org/10.2466/09.CP.4.22.

— (2020). *A Dream Survey*. URL: https://doi.org/10.6084/m9.figshare.13476879.v7.

— (2023). "When dreams come true: The relation of salient dreams to the future satisfaction of waking needs". In: *International Journal of Dream Research* 16.1, pp. 6–14. URL: https://doi.org/10.11588/ijodr.2023.1.89054.

Crevier, D. (1993). *The tumultuous history of the search for artificial intelligence*. Basic Books.

Damasio, Antonio R (1999). *The feeling of what happens: Body and emotion in the making of consciousness*. Houghton Mifflin Harcourt.

Darwin, Charles (1873). *The expression of the emotions in Man and animals*. Murray.

Delaney, Gayle MV (1997). *In Your Dreams: Falling, Flying, and Other Dream Themes: a New Kind of Dream Dictionary*. HarperSanFrancisco.

Dement, W. C. (1974). *Some Must Watch While Some Must Sleep*. W. H. Freeman.

Dement, W.C. et al. (2014). *Dement's Sleep and Dreams*. CreateSpace Independent Publishing Platform. ISBN: 9781502885883.

Domhoff, G William and Adam Schneider (2023). *Dream Bank.* Dreams used in this book were corrected for grammar and spelling. URL: http://www.dreambank.net.

Dorus, Elizabeth, Walter Dorus, and Allan Rechtschaffen (1971). "The incidence of novelty in dreams". In: *Archives of General Psychiatry* 25.4, pp. 364–368.

Dreamtime (2008). URL: https://www.pbs.org/video/wfyi-education-programs-dreamtime/.

Ekman, Paul (1999). "Basic emotions". In: *Handbook of cognition and emotion.* Wiley, pp. 45–60.

Emanuel, E. et al. (2000). "Understanding Economic and Other Burdens of Terminal Illness: The Experience of Patients and Their Caregivers". In: *Annals of Internal Medicine* 132.6, pp. 451–454.

Fagan, R. (1976). "Modeling how and why play works". In: *J. Bruner, A. Jolly, and K Sylva (Eds.), Play and its role in Development and Evolution.* Harmondsworth: Penguin, pp. 96–115.

Firestone, R. (1985). *The Fantasy Bond: Structure of Psychological Defenses.* Human Sciences Press.

Foulkes, David and Allan Rechtschaffen (1964). "Presleep determinants of dream content: Effects of two films". In: *Perceptual and Motor Skills* 19.3, pp. 983–1005.

Freud, Sigmund (1983). "The interpretation of dreams". In: *Literature and Psychoanalysis.* Columbia University Press, pp. 29–33.

Garfield, P. (2001). *The Universal Dream Key: The 12 Most Common Dream Themes From Around the World.* Cliff Street Books.

Gibson, W. (1969). *Dreams.* Grosset & Dunlap.

Goble, F. G. (1970). *The Third Force: The Psychology of Abraham Maslow.* Grossman Publishers.

Green, C. E. (1968). *Lucid Dreams.* Institute of Psychophysical Research.

Greetings People of Earth (2023). URL: https://www.thisamericanlife.org/803/greetings-people-of-earth.

Hall, C. S. and V. J. Nordby (1972). *The individual and his dreams.* New American Library.

Hall, Calvin Springer (1963). *Dreams of American College Students.* College women, late 1940s: #0441 (Code 064, Age 21, 10/21/47. University of Kansas.

Hampden-Turner, C. (1981). *Maps of the mind.* Collier.

Harford, Tim (2017). URL: https://www.bbc.com/news/business-38677721.

Harlow, Harry F. (1950). "Learning and satiation of response in intrinsically motivated complex puzzle performance by monkeys". In: *Journal of comparative and physiological psychology* 43.4, p. 289.

Hartmann, Ernest (1996). "Who develops PTSD nightmares and who doesn't". In: *Trauma and dreams*, pp. 100–113.

Hobson, J. Allan and Robert W. McCarley (1977). "The brain as a dream state generator: an activation-synthesis hypothesis of the dream process". In: *The American journal of psychiatry.*

Izard, Carroll E. (2007). "Basic emotions, natural kinds, emotion schemas, and a new paradigm". In: *Perspectives on psychological science* 2.3, pp. 260–280.

Izard, Carroll E. et al. (2015). "Emotions and developmental psychopathology". In: *Developmental Psychopathology: Volume One: Theory and Method*, pp. 244–292.

Johnson, Steven (2002). *Emergence: The connected lives of ants, brains, cities, and software.* Simon and Schuster.

Karni, Avi et al. (1994). "Dependence on REM sleep of overnight improvement of a perceptual skill". In: *Science* 265.5172, pp. 679–682.

Katz, J. (1994). *Dreams Are Your Truest Friends.* Simon and Schuster.

Keane, Andy J (1996). "The design of a satellite beam with enhanced vibration performance using genetic algorithm techniques". In: *Journal of the Acoustical Society of America* 99.4, pp. 2599–2603.

Khaleghi, Morteza and Karen Khaleghi (2011). *The Anatomy of Addiction: Overcoming the Triggers That Stand in the Way of Recovery.* St. Martin's Press.

Knudson, Roger M (2001). "Significant dreams: Bizarre or beautiful?" In: *Dreaming* 11.4, p. 167.

Kuriyama, Kenichi, Robert Stickgold, and Matthew P Walker (2004). "Sleep-dependent learning and motor-skill complexity". In: *Learning & memory* 11.6, pp. 705–713.

LaBerge, Stephen and Howard Rheingold (1990). *Exploring the world of lucid dreaming*. Ballantine Books.

Landrigan, Christopher P et al. (2004). "Effect of reducing interns' work hours on serious medical errors in intensive care units". In: *New England Journal of Medicine* 351.18, pp. 1838–1848.

Maslow, A. H. (1943). "Psychological Review". In: 50, pp. 370–396.

— (1954). *Motivation and Personality*. Harper & Row.

— (1969). "Theory Z. Journal of Transpersonal Psychology". In: 1.2, pp. 31–47.

— (1971). *The farther reaches of human nature.* Viking.

Maybruck, Patricia (1991). *Romantic Dreams: How to Enhance Your Intimate Relationship by Understanding and Sharing Your Dreams.* Pocket Books.

McClelland, James L and David E Rumelhart (1981). "An interactive activation model of context effects in letter perception: I. An account of basic findings". In: *Psychological review* 88.5, p. 375.

McCurdy, Harold Grier (1946). "The history of dream theory". In: *Psychological review* 53.4, p. 225.

Meadows, Donella H (2008). *Thinking in systems: A primer.* chelsea green publishing.

Meddis, Ray, AJD Pearson, and G Langford (1973). "An extreme case of healthy insomnia". In: *Electroencephalography and Clinical Neurophysiology* 35.2, pp. 213–214.

Miller, Stuart (2007). *Sometimes Not Just Curtains Fall Onstage*. URL: https://www.nytimes.com/2007/04/21/theater/21thud.html.

Morin, Roc (2015). "A conversation with Koko the gorilla". In: *The Atlantic* 28.

Muller, Karen Hagerman (1996). "Jasmine: Dreams in the psychotherapy of a rape survivor". In: *Trauma and dreams*. Harvard University Press, pp. 148–158.

Murrell, Stanley A., Suzanne Meeks, and James Walker (1991). "Protective functions of health and self-esteem against depression in older adults facing illness or bereavement". In: *Psychology and aging* 6.3, p. 352.

Nielsen, Tore A. (2000). "A review of mentation in REM and NREM sleep:"covert" REM sleep as a possible reconciliation of two opposing models". In: *Behavioral and Brain Sciences* 23.6, pp. 851–866.

Petit, C. (1998). "Touched by nature: Putting evolution to work on the assembly line". In: *U.S. News and World Report* 125.4, pp. 43–45.

Ramachandran, VS, Sandra Blakeslee, and Raymond J Dolan (1998). "Phantoms in the brain probing the mysteries of the human mind". In: *Nature* 396.6712, pp. 639–640.

Revonsuo, Antti (2000). "The reinterpretation of dreams: An evolutionary hypothesis of the function of dreaming". In: *Behavioral and brain sciences* 23.6, pp. 877–901.

Revonsuo, Antti, Jarno Tuominen, and Katja Valli (2015). "The avatars in the machine: Dreaming as a simulation of social reality". In: *Open mind*.

Robert, W. (1886). *Der traum als naturnothwendigkeit erklärt*. Hermann Seippel.

Rosen, Jeffrey B. (2004). "The neurobiology of conditioned and unconditioned fear: a neurobehavioral system analysis of the amygdala". In: *Behavioral and cognitive neuroscience reviews* 3.1, pp. 23–41.

Rosen, Melanie and John Sutton (2013). "Self-representation and perspectives in dreams". In: *Philosophy Compass* 8.11, pp. 1041–1053.

Rumelhart, David E. (2017). "Schemata: The building blocks of cognition". In: *Theoretical issues in reading comprehension*. Routledge, pp. 33–58.

Rycroft, C. (1979). *The Innocence of Dreams*. Random House.

Sagan, C. (1977). *The Dragons of Eden: Speculations on the Evolution of Human Intelligence*. Ballantine Books, p. 153.

Schoenberger, N. (1988). *Maintaining self-esteem in the face of behavioral presses for change: Support for the theory of self-verification*. Unpublished senior thesis, Princeton University.

Schultz, David, Carroll E. Izard, and George Bear (2004). "Children's emotion processing: Relations to emotionality and aggression". In: *Development and psychopathology* 16.2, pp. 371–387.

Sharot, T. (2011). *The optimism bias: A tour of the irrationally positive brain*. Pantheon/Random House.

Siegel, Alan (1996). "Dreams of firestorm survivors". In: *Trauma and dreams*. Harvard University Press, pp. 159–176.

Snyder, Mark and William B Swann (1978). "Hypothesis-testing processes in social interaction". In: *Journal of personality and social psychology* 36.11, p. 1202.

Software testing (2023). URL: https://en.wikipedia.org/wiki/Software_testing.

Spurr, P. (2001). *Dreams and Sexuality: Interpreting Your Sexual Dreams*. Sterling.

Starr, F. (2001). *Dream Themes: A Guide to Understanding Your Dreams*. Barnes & Noble Books.

Stoddard, Frederick J, David S Chedekel, and Laura Shakun (1996). "Dreams and nightmares of burned children". In: *Trauma and dreams*. Harvard University Press, pp. 25–45.

Sullivan, H. (1953). *Concepts of Modern Psychology*. W. W. Norton.

Taylor, J. (1992). *Where People Fly and Water Runs Uphill*. Warner Books.

Taylor, Jeremy (2009). *The wisdom of your dreams: Using dreams to tap into your unconscious and transform your life*. Penguin.

Tommaso, M. S. (2008). "Aesthetic value of paintings affects pain thresholds". In: *Consciousness and Cognition* 17.4, pp. 1152–1162.

Turing, Alan Mathison et al. (1936). "On computable numbers, with an application to the Entscheidungsproblem". In: *J. of Math* 58.345-363, p. 5.

Valli, Katja et al. (2006). "The effect of trauma on dream content–A field study of Palestinian children". In: *Dreaming* 16.2, p. 63.

Van de Castle, R. L. (1994). *Our Dreaming Mind*. Random House.

Verdone, Paul (1965). "Temporal reference of manifest dream content". In: *Perceptual and Motor Skills* 20.3, pp. 1253–1268.

Wagner, Ullrich et al. (2004). "Sleep inspires insight". In: *Nature* 427.6972, pp. 352–355.

Wamsley, Erin J. and R. Stickgold (2011). "Memory, sleep, and dreaming: Experiencing consolidation". In: *Sleep medicine clinics* 6.1, pp. 97–108.

Weisberg, Robert W (2006). *Creativity: Understanding innovation in problem solving, science, invention, and the arts*. John Wiley & Sons.

What Are Dreams (2009). URL: https://www.pbs.org/wgbh/nova/video/what-are-dreams.

Williams, Edwin A, William A Crossley, and Thomas J Lang (2001). "Average and maximum revisit time trade studies for satellite constellations using a multiobjective genetic algorithm". In: *The Journal of the astronautical sciences* 49, pp. 385–400.

Wilmer, Harry A (1996). "The healing nightmare: War dreams of Vietnam veterans". In: *Trauma and dreams*. Harvard University Press, pp. 85–89.

Winerman, Lea (2006). "Let's sleep on it". In: *Monitor on Psychology* 37.1, p. 58.

Winson, J. (1990). "The Meaning of Dreams". In: 262, pp. 86–96.

Wood, James M et al. (1992). "Effects of the 1989 San Francisco earthquake on frequency and content of nightmares". In: *Journal of abnormal psychology* 101.2, p. 219.

Xie, Lulu et al. (2013). "Sleep drives metabolite clearance from the adult brain". In: *science* 342.6156, pp. 373–377.

Young, J. E., J. S. Klosko, and M. E. Weishaar (2003). *Schema therapy*. Guilford.

Index

INDEX

Made in the USA
Monee, IL
08 November 2026

67378332R10204